ICS 91.080.40
P 25

Standard of Ministry of Water Resources of the People's Republic of China

SL 319—2005
Replace SL 21—78

Design Standard for Concrete Gravity Dams

Issued on July 21, 2005   Implemented on November 1, 2005

Yellow River Conservancy Press
· Zhengzhou ·

图书在版编目(CIP)数据

混凝土重力坝设计规范=Design standard for concrete gravity dams:英文/中华人民共和国水利部发布.—郑州:黄河水利出版社,2013.12
ISBN 978-7-5509-0533-7

Ⅰ.混… Ⅱ.中… Ⅲ.混凝土坝-重力坝-设计规范-中国-英文 Ⅳ.①TV642.3-65

中国版本图书馆CIP数据核字(2013)第202198号

出 版 社:黄河水利出版社
　　　　　地址:河南省郑州市顺河路黄委会综合楼14层　邮政编码:450003
发行单位:黄河水利出版社
　　　　　发行部电话:0371-66026940、66020550、66028024、66022620(传真)
　　　　　E-mail:hhslcbs@126.com
承印单位:黄河水利委员会印刷厂
开本:850 mm×1 168 mm　1/32
印张:4.75
字数:230千字　　　　　　　　　　印数:1—1 000
版次:2013年12月第1版　　　　　　印次:2013年12月第1次印刷
定价:150.00元

Introduction to English Version

Department of International Cooperation, Science and Technology of Ministry of Water Resources, P. R. China has the mandate of managing the formulation and revision of water technology standards in China.

Translation of this English version of standard was organized by Department of International Cooperation, Science and Technology of Ministry of Water Resources, P. R. China in accordance with due procedures and regulations applicable in the country.

This English version of standard is identical to its Chinese original *Design Standard for Concrete Gravity Dams* (SL 319—2005), which was formulated and revised under the auspices of Department of International Cooperation, Science and Technology of Ministry of Water Resources, P. R. China.

Translation of this standard was undertaken by China Water Resources Beifang Investigation, Design and Research Co. Ltd.

Translation team includes Du Leigong, Chen Shaosong, Yang Kete, Yang Haiyan, Chen Honglian, Zhang Baorui, Yin Zize, Ding Xiuxia, and Gao Ying.

This standard was reviewed by Du Chongjiang.

Department of International Cooperation, Science and Technology
Ministry of Water Resources, P. R. China

Foreword

Design Standard for Concrete Gravity Dams was initially issued in 1978 and amended in 1984. According to File No. 1 [2001] issued by Water Resources and Hydropower Planning and Design Administration of the Ministry of Water Resources and pursuant to *Specifications for Drafting of Technical Standards of Water Resources* (SL 1—2002), the previous edition of *Design Standard for Concrete Gravity Dams* (SDJ 21—78) is amended.

This standard comprises 10 chapters with the main contents as follows:

—Layout of gravity dam.

—Configuration selection of solid gravity dam, slotted gravity dam, and hollow gravity dam; and structural layout of water releasing works.

—Hydraulic design for flood discharge, energy dissipation and scour protection.

—Calculation of loads, dam stress and dam stability, and related criteria.

—Dam foundation treatment, including excavation, consolidation grouting, seepage control and drainage, treatment of Karst cave and fault fracture zones.

—Dam structure, dam material, and structures of dam crest, gallery, joints, water seals, and drainages.

—Temperature control criteria and cracking prevention measures.

—Dam safety monitoring.

The following is the summary of the significant technical differences between the previous edition (SDJ 21—78) and this revision:

—Add flood control standard for energy dissipation works.

—Add design of new types of energy dissipating devices; supplement design related to atomization during flood discharging.

—Add contents of consolidation grouting without concrete cover or with thin concrete cover.

—Add contents of seepage control in Karst areas.

—Supplement the factor of autogenous volumetric change of concrete in design of concrete cracking prevention measures; add that the factors of elastic deformation modulus of foundation and linear expansion coefficient of concrete shall be taken into account when deciding the allowable foundation temperature difference; add provisions on temperature difference between the interior and surface of concrete; add calculation method of concrete surface insulation.

—Rename "observation design" as "safety monitoring design"; modify classification and definition of different tasks of safety monitoring; specify clearly the scope of safety monitoring; add principles for safety monitoring design; adjust special monitoring items; add installation requirements for major monitoring devices.

—A separate standard has been drafted duly for design of RCC gravity dams, thus requirements for RCC gravity dam design is no more within the coverage of this standard.

The provisions printed in boldface are compulsory and must be enforced, including Clause 6.3.2, Clause 6.3.4, Clause 6.3.10, Clause 6.4.1, Clause 7.4.5, Item 1 of Clause 10.1.1, Item 5 of Clause 10.1.4, and Item 1 and Item 2 of Clause 10.2.2.

This document replaces the standard under No. SDJ 21—78.

This standard was approved by Ministry of Water Resources of the People's Republic of China.

This standard was initiated by Water Resources and Hydropower Planning and Design Administration of Ministry of Water Resources.

This standard will be explained by Water Resources and Hydropower Planning and Design General Institute.

This standard was chiefly drafted by Changjiang Institute of Survey, Planning, Design and Research.

This standard is published and distributed by China Water Power Press.

Chief drafters of this standard are Xu Linxiang, Wang Xiaomao, Wang Youyang, Chen Jitang, Liao Renqiang, Hu Jinhua, Guo Yanyang, Wang Anhua, Wang Qingyuan, Fan Wuyi, Lei Xingshun, Zhang Zhiyong, Zhou Heqing, Gao Dashui, Xiang Guanghong, Fan Luqi.

The technical responsible person of this standard review conference is Shen Fengsheng.

The format examiner of this standard is Cao Yang.

Contents

1 **General Provisions** .... (1)
2 **Terms and Notations** .... (2)
    2.1 Terms .... (2)
    2.2 Notations .... (2)
3 **Layout of Dam** .... (4)
4 **Configuration of Dam** .... (7)
    4.1 General .... (7)
    4.2 Non-overflow Dam Sections .... (7)
    4.3 Overflow Dam Sections .... (9)
    4.4 Outlet Works .... (10)
5 **Hydraulic Design of Water Releasing Works** .... (13)
    5.1 General .... (13)
    5.2 Calculation of Discharge Capacity and Energy Dissipation .... (15)
    5.3 Design of Cavitation Prevention in High Velocity Flow Area .... (16)
    5.4 Design of Energy Dissipating and Scouring Resistant Devices .... (17)
6 **Design of Dam Cross Section** .... (20)
    6.1 Loads and Loading Combinations .... (20)
    6.2 Design Principles .... (22)
    6.3 Stress Computation .... (23)
    6.4 Calculation of Dam Stability against Sliding .... (26)
    6.5 Structural Design of Gate Piers on Overflow Dam Section .... (29)

7	**Foundation Treatment**	(31)
	7.1 General	(31)
	7.2 Foundation Excavation	(31)
	7.3 Consolidation Grouting of Dam Foundation	(33)
	7.4 Seepage Control and Drainage of Foundation	(34)
	7.5 Treatment of Fault Fracture Zones and Weak Structural Planes	(38)
	7.6 Seepage Control in Karst Areas	(40)
8	**Structural Arrangement of Dam Body**	(42)
	8.1 Dam Crest	(42)
	8.2 Galleries and Access Passages in Dam	(43)
	8.3 Joints in Dam	(45)
	8.4 Waterstop and Drainage	(46)
	8.5 Concrete Materials and Dam Zoning	(48)
9	**Temperature Control and Crack Prevention**	(53)
	9.1 General	(53)
	9.2 Temperature Control Criteria	(54)
	9.3 Temperature Control and Cracking Prevention Measures	(56)
10	**Safety Monitoring Design**	(60)
	10.1 General	(60)
	10.2 Monitoring Items and Highlights for Instrument Layout	(62)
Annex A	**Formulas for Hydraulic Design**	(69)
	A.1 Weir Crest Shape, Pressure on Crest Surface and Bucket Radius	(69)
	A.2 Outline of Outlet Works through Dam Body	(75)
	A.3 Calculation of Discharge Capacity and Aerated Water Depth	(79)
	A.4 Hydraulic Characteristics of Flip Bucket Energy	

	Dissipation	(81)
A.5	Hydraulic Characteristics of Energy Dissipation of Hydraulic-jump Type Stilling Basin	(84)
A.6	Prevention of Cavitation	(85)
Annex B	**Formulas for Load Calculations**	(89)
B.1	Hydrostatic Pressure at a Point on Dam Surface	(89)
B.2	Silt Pressure	(89)
B.3	Uplift Pressure	(89)
B.4	Ice Pressure	(93)
B.5	Pressure on the Bucket by Centrifugal Flow	(95)
B.6	Wave Pressure	(96)
Annex C	**Formulas for Stress Calculations of Solid Gravity Dams**	(102)
C.1	Vertical Normal Stresses at Upstream and Downstream Faces	(102)
C.2	Shear Stresses at Upstream and Downstream Faces	(102)
C.3	Horizontal Normal Stresses at Upstream and Downstream Faces	(103)
C.4	Principal Stresses at Upstream and Downstream Faces	(103)
Annex D	**Engineering Geological Classification and Mechanical Parameters of Rock Mass of Dam Foundation**	(105)
Annex E	**Calculations of Stability against Sliding along Potential Failure Planes within Foundation**	(112)
Annex F	**Calculation of Temperature and Thermal Stresses of Dam during Construction**	(115)
F.1	Calculation of Concrete Temperature	(115)
F.2	Calculation of Temperature Reducing of Concrete by	

    Pipe Cooling ··· (124)
 F.3 Thermal Insulation of Concrete Surface ············ (127)
 F.4 Thermal Stresses ··· (130)
Expression of Provisions ··· (136)

1 General Provisions

1.0.1 This standard is formulated to standardize design of concrete gravity dams, so that the design could meet the requirements of safety, economy, state-of-art technology and reliable quality, keeping pace with the development of concrete gravity dam construction.

1.0.2 This standard is applicable to the design of Class 1, Class 2 and Class 3 concrete gravity dams on rock foundation for large and medium water resources and hydroelectric projects, and it may also be used as a guide for designing Class 4 and Class 5 concrete gravity dams.

For concrete gravity dams over 200 m high or those of particular significance, special studies shall be conducted on the associated special problems.

1.0.3 Concrete gravity dams are classified into low dam, medium dam and high dam by the structural height. Low dams are those below 30 m high, medium dams are those from 30 m to 70 m high (including 30 m and 70 m), and high dams are those over 70 m high.

1.0.4 The standards referenced in this standard mainly include:

Standard for Flood Control (GB 50201);

Standard for Classification and Flood Control of Water Resources and Hydroelectric Projects (SL 252);

Design Specification for Power Intake (SL 285);

Specifications for Seismic Design of Hydraulic Structures (SL 203);

Design Specifications of Hydraulic Structures against Ice and Freezing Action (SL 211);

Hydraulic and Hydroelectric Engineering Specification for Design of Steel Gate (SL 74);

Design Code for Hydraulic Concrete Structures (SL/T 191);

Technical Specification for Cement Grouting Construction of Hydraulic Structures (SL 62).

1.0.5 Besides this Standard, design of a concrete gravity dam shall also conform to other relevant national and ministerial standards of the People's Republic of China.

2 Terms and Notations

2.1 Terms

2.1.1 Dam height

Vertical distance between the dam crest and the lowest point of the dam base (exclusive of localized deep trench, pit, or cavern).

2.1.2 Solid concrete gravity dam

A gravity dam fully filled and placed with concrete except some small cavities.

2.1.3 Hollow concrete gravity dam

A concrete gravity dam in which a hollow space is provided with its longer dimension in the direction parallel to the dam axis.

2.1.4 Slotted concrete gravity dam

A concrete gravity dam of which the middle parts of the transverse joints are widened to form hollow spaces between adjacent monoliths.

2.1.5 Flaring pier

A type of spillway pier with a flared end.

2.1.6 Combined energy dissipation

Energy dissipation by combined use of more than one type of energy dissipators, including the combined flaring pier with flip bucket, or the combined flaring pier with hydraulic jump stilling basin, or the combined flaring pier with roller bucket type, or others.

2.1.7 Skew type flip bucket

A type of flip bucket with a distorted bottom surface and a bucket lip with varied height deflecting out flow laterally at certain angle with the incoming flow direction.

2.1.8 Slit type flip bucket

A type of flip bucket with sharply contracted sidewalls at the exit of the chute to form a narrow passage.

2.2 Notations

2.2.1 Material properties

γ_w—unit weight of water;
C_w—specific heat of water;
α_C—thermal diffusivity of concrete;
λ_C—thermal conductivity of concrete;
α—coefficient of linear thermal expansion of concrete;
E_C—elastic modulus of concrete;
μ—Poisson's ratio of concrete;
γ_C—unit weight of concrete;
f'—friction coefficient of shear-friction strength;
C'—cohesion of shear-friction strength;
f—friction coefficient of friction strength;
E_R—deformation modulus of foundation rock.

2.2.2 Geometric features

B—net width of spillway weir;
A_k—area at exit of outlet orifice.

2.2.3 Calculation parameters

K'—safety factor of stability against sliding calculated using shear-friction strength;
K—safety factor of stability against sliding calculated using friction strength.

2.2.4 Calculation parameters

v—flow velocity;
Q—discharge;
H_d—design head on spillway crest;
t_k—water depth in scour hole;
h_b—depth of fluctuating or aerated water flow;
P_m—intensity of pulsating pressure;
P_d—impact force acting on the stilling basin baffle piers.

2.2.5 Calculation coefficients

σ_K—cavitation number;
Fr—Froude number;
m—discharge coefficient of spillway;
δ_s—submergence coefficient;
φ—coefficient of flow velocity;
K_p—coefficient of stress relaxation due to concrete creep.

3 Layout of Dam

3.0.1 The dam layout shall be given comprehensive considerations in combination with the project layout, on the rational arrangement of various component structures for flood release, water supply, power generation, irrigation, navigation, sediment flushing, floating debris removal, fish passage, and others so as to prevent mutual interference. Top priority should be given to the layout of flood releasing structure to ensure that the discharged flow will not undermine the dam foundation or other structures, and that the flow regime and river erosion and deposition will not adversely influence the serviceability of other structures.

3.0.2 The total length, the bay number, the type and dimension of weir, and the weir crest of the overflow dam section shall be determined by comprehensive comparison of the following factors (free overflow spillway, with its large discharge capacity, should generally be given preference as the first choice):

  a Requirements of reservoir operation, flood releasing and floating debris removal.

  b Topographical and geological conditions at the dam site, and erosion resistance of the downstream riverbed and banks.

  c Requirements of downstream water depth and energy dissipation.

  d Dam monolith arrangement and interaction with adjacent structures.

  e Type, working condition and operating rule of the gates.

3.0.3 The type of energy dissipation and scour resistance shall be rationally selected based on the dam height, the topographical and geological conditions of the downstream riverbed and banks, and the fluctuation of the downstream water depth in combination with other requirements for handling the floating debris, ice flushing, etc.

3.0.4 Outlet works shall be provided in the dam body to fulfill the requirements of the following functions:

a Flood releasing.

b Reservoir drawdown or emptying, design earthquake intensity higher than Ⅷ, or extremely complicated geological conditions of the dam foundation.

c Water supply to downstream area.

d Sediment flushing.

e Temporary flood releasing devices during construction to be incorporated into permanent flood releasing works during operation.

3.0.5 The type, location, elevation, number and size of the outlet works shall be selected by considering the following factors:

a Layout conditions: in a narrow river channel, the outlet works should be incorporated with the overflow dam sections, and the energy dissipators shall be jointly considered with the overflow dam. In a wide river channel, the outlet works may be separately arranged. The outlet works for sediment flushing shall be arranged close to the intake for power (or for irrigation and for water supply) and close to the headwork of navigation lock, etc. so that the sediment flushing flow pattern shall not influence the normal operation of these nearby structures.

b Operation conditions: the flow discharge, discharge duration, maintenance, sediment flushing, floating debris flushing, etc.

c Construction conditions: the effects of different positions of the outlet works on the construction method and schedule; the requirements for flood releasing and for water supply to downstream area during construction.

d Gate operating conditions: the hoisting equipment, the dam structural strength, etc.

3.0.6 The construction diversion works (bottom outlets, gaps and slotted construction blocks) to be arranged in the dam body shall be determined based on the river diversion scheme, topography, geology, hydrology, etc. The following requirements shall be considered for their layout (proper design and construction measures shall be taken for plugging the river diversion works):

a To be capable of releasing the design flow during construction.

b To be capable of combining with the permanent water releasing works.

c To be navigable during construction in a navigable river, or to adopt other measures to meet navigation requirements.

d To be capable of passing floating ice and other floating debris, when required.

e To be capable of preventing damage to the permanent structures or influencing the construction schedule during flood releasing.

f To be convenient in construction, reliable in operation and easy for backfilling and plugging.

3.0.7 The elevations of water intake for agricultural, industrial, and municipal water supply shall be determined according to the requirements for the water level, flow rate and water amount during the water supply period. If necessary, multi-level intake orifices should be provided according to water temperature and reservoir sedimentation.

The intake elevation of power conduits within the dam body shall be determined based on the hydro energy design and reservoir sedimentation.

The arrangement of various intake structures shall also comply with *Design Specification for Power Intake* (SL 285).

Preferably, the inlet and outlet of dam-passing structures should be located far away from those of flood releasing structures.

3.0.8 For the gravity dam layout of a large project, overall hydraulic model tests shall be performed to verify whether the flow patterns and the river erosion and deposition conditions during construction and operation can meet the operation requirements of all component structures.

Overall hydraulic model tests are also desirable for medium-sized hydro projects.

4 Configuration of Dam

4.1 General

4.1.1 The dam configuration shall be determined through comprehensive technical and economic comparison, according to loading conditions of the dam, and according to topography, geology, hydrology, meteorology, construction materials, and construction period at the dam site.

4.1.2 The upstream faces of all dam sections are desirable to be consistent with each other. The waterstops in the upstream part of the transverse joints at both sides of a dam monolith should be installed symmetrically. If adjustment is needed, the transition should be gradual.

4.1.3 The dam crest elevation shall be determined in accordance with Clause 8.1.1 of this standard.

4.1.4 A concrete gravity dam built in a seismically active region shall meet the requirements prescribed in *Specifications for Seismic Design of Hydraulic Structures* (SL 203).

4.1.5 A concrete gravity dam built in a cold region shall meet the requirements prescribed in *Design Specification of Hydraulic Structures against Ice and Freezing Action* (SL 211).

4.1.6 A slotted concrete gravity dam, massive-head buttress dam, or hollow concrete gravity dam, other than a solid concrete gravity dam, may also be adopted if it is deemed feasible through technical and economic comparison.

4.2 Non-overflow Dam Sections

4.2.1 The basic cross-section of a non-overflow dam section is essentially a triangle, with its apex preferably near to the dam crest. The upper portion of the basic cross-section is shaped to accommodate the dam crest structure.

4.2.2 The upstream face of a dam may be vertical, sloped, or broken-line face. For a solid gravity dam, the upstream face should be sloped between 1 : 0.2 and 1 : 0. If the upstream slope has a turn, the elevations at the turn point shall be optimally determined with consideration of the arrangement of power intake, outlet works, etc., as well as the slope of the downstream face.

The downstream face may have one slope or more slopes, which shall be determined by the stability and stress requirements in conjunction with the upstream slope. The downstream face is normally sloped between 1 : 0.6 and 1 : 0.8. For a monolithic concrete gravity dam, in which the transverse joints are keyed and grouted, the dam slope may be selected by taking into account the combined action between the neighboring dam monoliths.

4.2.3 The upstream face of a slotted gravity dam should have a gentler slope comparing to a solid gravity dam. The width of the slot may range from 20% to 40% of the width of the dam monolith.

When power conduits, water release outlets, or diversion bottom outlets are to be provided in the dam body, the dam structure and slot arrangement of that dam section shall be determined through verification.

4.2.4 The minimum thickness of the upstream head of a slotted gravity dam may be taken as 0.07-0.1 times the reservoir water head at the point considered, and shall not be less than 3 m. The minimum thickness of the downstream end part of the slotted dam is better to be more than 2 m, and shall be increased properly in the case of dams built in cold regions. The dimension of the upstream dam head shall be designed by considering the factors as below:

   a Stress state in the dam head.

   b Arrangement of seepage control and waterstop system at the upstream face.

   c Arrangement of galleries for curtain grouting, drainage, and inspection.

   d Other requirements.

4.2.5 The slots of a slotted gravity dam shall not extend up to the dam crest. Sufficient length of transition shall be provided at the upstream, downstream and upper portions of the slot. The transition slopes (i. e. the ratio of the length parallel to the dam axis to the length perpendicular to the dam axis) of the slot in the horizontal section may be taken between 1 : 2.0 and 1 : 1.5 for the upstream portion, and between 1 : 1.5 and 1 : 1.0 for the downstream portion. The transition slopes (i. e. the ratio of the vertical height to the horizontal length) of the upper portion of the slot in the vertical section may be taken between 1.5 : 1 and 2.0 : 1. The top elevation of the slot should be determined by considering the stability and stress requirements, as well as the slopes of upstream and downstream faces, the width of the slot, etc.

4.3 Overflow Dam Sections

4.3.1 The profile of the weir crest surface may be defined by an exponential curve if a free flow spillway is used and by a parabolic curve if orifice flow is used with breast wall to retain water. Annex A.1 shows the details for determining the profile of the weir.

Other type of weir profiles may also be adopted, provided that they are verified by careful studies and tests.

4.3.2 When the gates are fully open to discharge frequent floods under local atmospheric conditions, negative pressure should not occur near the weir crest. When the gates are partially open, small negative pressure is allowed to occur only after justification. When the gates are fully open to discharge the design floods, the negative pressure shall not exceed 3×9.81 kPa. When the gates are fully open to discharge the maximum floods, the negative pressure shall not exceed 6×9.81 kPa.

4.3.3 The type of gate slots shall be carefully selected to prevent occurrence of excessive negative pressure, which may cause cavitation damage to the gate slots.

4.3.4 The reverse curve of the overflow sections shall be selected in combination with the downstream energy dissipators, see Annex A.1.

4.3.5 The type and size of gate piers shall meet the requirements of structural arrangements and hydraulic conditions. When a vertical-lift gate is adopted, the piers shall be of adequate thickness at the gate slots so as to meet the strength requirements.

4.3.6 For a large project, hydraulic model tests shall be conducted for the overflow dam sections to verify the profile of the weir surface, the type of the gate piers, the gate slots, the pressure on the weir surface, the discharge capacity, the radius of reverse curve, etc. For a medium project, such hydraulic model test should be conducted. For projects under simple hydraulic conditions, the above design features may be determined by hydraulic calculation with reference to the experiences of other similar projects.

4.3.7 When the overflow spillways are required to pass ice, the size of spillway bays shall be determined to fit the ice conditions. The depth of flow over the crest is desirable to be greater than the maximum ice thickness. The ice blocks can freely pass through the dam without causing damage to the downstream facilities. Guide walls and bank revetments shall be provided at the downstream area. The nose of the gate piers in the shape of a sharp angle is preferable and should be determined by tests if necessary.

4.3.8 The gates provided on the overflow sections shall comply with the requirements in *Hydraulic and Hydroelectric Engineering Specification for Design of Steel Gate* (SL 74).

4.4 Outlet Works

4.4.1 Outlet works may be arranged in the lower part of the overflow dam sections or in one special dam section, and energy dissipating devices shall be provided.

4.4.2 For outlet works provided in the dam, the alternate occurrence of pressure flow and free flow shall be avoided inside the opening.

4.4.3 Either free flow or pressure flow outlet works through the dam may be adopted in the design.

4.4.4 A free flow opening consists of a pressure flow portion and a free flow portion.

The pressure flow portion comprises the entrance, gate slot and constriction segments. This pressure flow portion shall be shaped to ensure positive pressure at all flow discharges, to have uniform change in its cross section and to attain high discharge capacity. Service gates are provided at the downstream end of the pressure flow portion, with emergency or bulkhead gates provided at the upstream side. The design of the profile of the pressure flow portion is given in Annex A.2.

The free flow portion shall have an adequate clearance below the top of the opening. For the straight-line sections, the clearance at its maximum discharge may be determined to be: 30%-50% of the non-aerated flow depth for a rectangular cross-section (here the clearance is the height between the roof and the water surface); 20%-30% of the non-aerated flow depth for a cross-section with circular arch roof, (here the clearance is defined as the height from the arch spring line to the water surface). The clearance may be slightly greater for a cross section with an elliptical roof than the value required for the cross-section with circular arch roof to ensure no submergence of the roof.

The exit of the free flow portion should be higher than the tailwater level to avoid occurrence of hydraulic jump in this segment.

When the flow velocity in the free flow portion is rather high, air entrainment measures shall be taken to mitigate the cavitation damage.

4.4.5 The free flow opening is preferably of a straight alignment in plan; in case it has to be curved, careful study shall be carried out and verified by hydraulic model tests.

4.4.6 The requirements for the entrance segment arrangement of a pressure flow opening are basically the same as those for that of a free flow opening. Downstream of the entrance segment, the slot segment for an emergency gate should be arranged and then followed by a segment with a flat floor or floor of a gentle slope of less than 1 : 10. The service gate is provided at the outlet end part of the pressure flow opening, upstream of which is the constriction segment. The cross-section of pressure flow opening may be either rectangular or circular in shape. The design of the profile of a pressure flow opening is given in Annex A.2.

4.4.7 The design of the gate and hoisting equipment of an outlet works in the dam body shall meet the following requirements:

The service gate for free flow opening may be either a radial gate or a vertical-lift gate, but the emergency gate will be a vertical-lift gate. The hoist chamber for the radial gate should arranged in the dam body, and it should be located on the dam crest if the dam is of medium height. The hoist chamber for the vertical-lift gate is preferably arranged on the dam crest. When the hoist chamber is located in the dam, air ventilation, damp proofing, and heating devices shall be provided.

The service gate for pressure flow opening may be a radial gate, vertical-lift gate, Howell Bunger valve, or other appropriate types of gate or valve. When the pressure flow opening is not provided with steel liner, the emergency gate will be used to retain water during non-flood season preferably.

4.4.8 When the area downstream of the gate of an outlet opening in the dam cannot be adequately aerated, an air vent shall be provided in the roof of the outlet opening immediately downstream of the gate. The upper end of the air vent shall be separated from the hoist chamber with protection facilities. The air vent shall be designed in conformity to the relevant requirements in *Hydraulic and Hydroelectric Engineering Specification for Design of Steel Gate* (SL 74).

4.4.9 In the case of a high dam with an inside outlet works, where the hydraulic conditions are complicated, hydraulic model tests shall be performed, and verified by vacuum tank test, if desired.

4.4.10 When it is unavoidable that flood needs to be released through the low-level river diversion outlet and through the upper-level water releasing devices at the same time, due consideration shall be given to the unfavorable situation that the outflow from the low level river diversion outlet may be obstructed. Measures shall be taken to prevent cavitation damage.

4.4.11 The lining of an outlet opening through the dam shall be determined on the basis of hydraulic conditions, opening size, sediment concentration features of the flow, operating conditions, etc. The pressure flow openings as well as the pressure flow portions of free flow openings, where the internal water pressure is high, should be lined with steel or high performance concrete. The steel liner shall be firmly bonded to the surrounding concrete.

5 Hydraulic Design of Water Releasing Works

5.1 General

5.1.1 Hydraulic design of water releasing works includes:
    a Calculation of discharge capacity.
    b Design of the connection of outflow with downstream flow and the energy dissipation and scouring resistance facilities.
    c Hydraulic design in relation to the high velocity water flow.
    d Other relevant hydraulic designs.

5.1.2 The design flood standard for hydraulic design of water releasing works and energy dissipating structures shall conform to the requirements in *Standard for Flood Control* (GB 50201) and in *Standard for Classification and Flood Control of Water Resources and Hydroelectric Projects* (SL 252).

5.1.3 Hydraulic computation of water releasing works may use the equations given in Annex A.

5.1.4 In addition to Clause 3.0.3 of this standard, the design of energy dissipation and scouring resistance of water releasing works shall meet the following requirements:
    a The energy dissipators shall be highly effective, structurally reliable, and highly resistant to cavitation and abrasion, and shall be capable of preventing the erosion of dam foundation and downstream riverbanks, and ensuring the safety of the dam and the appurtenant structures.
    b The selected energy dissipator shall achieve good effectiveness in energy dissipation in releasing the design flood and all other minor floods, especially the frequently occurring floods. For a flood exceeding the design standards for the energy dissipator, localized damages to the energy dissipating and scouring protection works are permissible, provided the safety of the dam will not be jeopardized, the long-term operation of the project will not be affected and the damaged portion can be easily repaired.

c Provisions shall be made for the submerged stilling basins or roller buckets to meet the maintenance requirements for dewatering and repairing when needed in the operation period.

5.1.5 Flip buckets are applicable to high or medium dams built on sound rock. They may be used for low dams only after justification.

The flip bucket is generally not applicable to the condition when a gently dipping weak plane exists in the dam foundation and extends further downstream, which may be intersected by the scour hole to form an overhanging plane, consequently, jeopardizing the dam foundation stability, or sloughing the river banks. If the flip bucket is to be adopted in such case, special protective measures shall be provided.

5.1.6 The hydraulic jump stilling basin is applicable to medium or low dams, or to dams built on river channel with weak bedrock. If it is to be used for a high dam, special verification shall be conducted, but it is unsuitable for passing floating debris or ice.

5.1.7 "Surface-type energy dissipater" is applicable to medium or low dams, having relatively low water head, straight river channel, steady water level, deep tailwater, high scouring-resistant river bed and banks stretching to a certain range, and being capable of flushing floating debris or ice.

5.1.8 The roller bucket energy dissipator is suitable for the river channel with deep tailwater and with riverbed and banks having certain resistance to scouring.

5.1.9 The combined energy dissipators are applicable to high or medium dams with large flood discharge, relatively narrow river channel, and fairly poor geological conditions in the downstream area or applicable to the condition whenever a single-type energy dissipator is not economical. The combined energy dissipators shall be verified by hydraulic model tests.

5.1.10 The gates of water releasing structures should be hoisted or closed in a synchronous, symmetrical and uniform manner in order to stabilize the flow pattern. The operating program of the gates shall be prepared.

5.1.11 For a large project or a high dam, the design of the water releasing structures shall be verified by hydraulic model tests. For a medium-sized hydro project, the hydraulic model tests may be needed to verify the design. For some medium-sized projects with simple hydraulic conditions, the water releasing structures may be designed through hydraulic computation with reference to the experiences of other similar projects.

5.2 Calculation of Discharge Capacity and Energy Dissipation

5.2.1 The discharge capacity of an overflow dam and an outlet works may be calculated in accordance with Annex A.3.

5.2.2 The calculation of water surface profile on a spillway shall consider the effect of fluctuation and aeration when the Froude number $Fr > 2$. The calculation may refer to the equations in Annex A.3. The top of sidewalls or training walls of a spillway chute shall be 0.5-1.5 m higher than the calculated water surface profile.

5.2.3 The design of a flip bucket energy dissipator shall include hydraulic calculation for all various discharge flows. The jet trajectory distance and the maximum depth of the scour hole may be calculated using the equations given in Annex A.4.

5.2.4 The design of a hydraulic jump stilling basin shall include hydraulic calculation for all various discharge flows, to determine the elevation of apron floor, length of basin, thickness of apron, and the degree of tailwater submergence.

5.2.5 The apron length may be calculated by using equations in Annex A.5 considering the hydraulic characteristics of with or without the appurtenant structures (baffle piers and end sills). When no appurtenant structures are provided on the apron, the tailwater depth may be taken as 1.05-1.10 times the water depth after the jump.

5.2.6 The time-averaged water pressure on the apron may be determined as follows:

a When the apron is horizontal, the time-averaged water pressure acting on the apron may be approximately taken as the water depth at the computational cross section.

b When the apron is not provided with baffle piers and end sill as appurtenant structures, a straight line connecting the water surfaces at the starting and ending points of the hydraulic jump may be deemed as an approximate water-surface profile.

c When the apron is provided with baffle piers, the water pressure downstream of the piers may be calculated by using the water depth after the jump, while those upstream of the piers may use a half of the water depth after the jump.

5.2.7 The pulsating pressures acting on the bucket lip, on the roof of overflow powerhouse, on the apron, etc., and the impact forces acting on the baffle piers and end sills, may be calculated by using the equations given in Annex A.5.

5.3 Design of Cavitation Prevention in High Velocity Flow Area

5.3.1 In high-velocity flow areas of water releasing works, attention shall be paid to potential cavitation damages to the following parts or area:

a Entrance and exit parts, gate slot, bend section, and locations where flow boundary changes abruptly.

b Reverse curve portion and its vicinity.

c Special-shaped bucket lip, flow dividing pier.

d Surface of overflow dam and walls of outlet works, where the flow velocity exceeds 20 m/s.

5.3.2 The cavitation number (σ) at different locations in the high velocity flow area should be higher than the incipient cavitation number at the same location. The cavitation number may be estimated by using the equations in Annex A.6.

5.3.3 For the locations prone to cavitation damages, the following preventive measures shall be adopted:

a Select rational outlines and dimensions of the structures.

b Control the unevenness of flow passage surfaces as per Table A.6.2 in Annex A.

c Provide aeration devices as described in Clause A.6.3 in Annex A. The criteria for controlling the unevenness of flow passage surfaces may be properly less stringent at the aerated areas.

d Use surface protection materials with high performance of cavitation resistance.

e Establish rational operation program.

5.3.4 Preferably, aeration devices should be provided for water release works with flow velocity exceeding 30 m/s. For works of particularly important or for works with flow velocity exceeding 35 m/s, vacuum tank model tests should be performed to determine the cavitation prevention measures.

5.3.5 For water release works on silt-laden rivers, consideration shall be given to the interaction between abrasion and cavitation of the silt-laden flow at high velocity.

5.4 Design of Energy Dissipating and Scouring Resistant Devices

5.4.1 The lip of flip bucket has different types, e. g. solid type, dentate type, slit type, skew type and others. Which type to be selected shall be determined after comparative study. Normally, the minimum bucket lip elevation should be higher than the downstream water level corresponding to the design flood as stipulated in Clause 5.1.2; but it may be slightly lower than the maximum tailwater level. The lip angle should be determined through comparison.

a If adentate bucket lip is adopted and the mean flow velocity at the bucket lip exceeds 16 m/s, rational selection shall be made on the reverse curve radius, the difference in trajectory angles, the ratio of low teeth width to high teeth width, and the height difference between the high and low teeth. A straight transition section may be provided between the reverse curve of the bucket and the bucket lip for improving the flow regime. The difference of the trajectory angles between the high teeth and the low teeth of a dentate lip may range from 5° to 10°. The ratio of the width of high teeth to the width of low teeth should be greater than 1.0. It is suitable for the teeth to have a height differential of around 1.5 m. Air vent holes are advised to be provided on the sides of the high teeth, and the edges on the top of the high teeth should be rounded.

b Slit-type bucket lip is applicable to high-head spillway and low-level outlets in narrow river channel. The exit opening may have a rectangular, trapezoidal, Y-shaped, V-shaped or even asymmetrically shaped cross section. The cross-section, trajectory angle, contraction ratio, and length-to-width ratio of the slit-type bucket lip should be determined through comparative study on model test.

   c For the skew-type bucket lip, the lateral deflecting angle and jet nappe impinging point in the downstream water surface shall be determined through hydraulic model tests.

5.4.2 The safe jet trajectory distance shall be determined by ensuring the stability of the foundation bedrock at the dam toe. The distance between the lowest point of the scour hole and the dam toe shall exceed 2.5 times the depth of the scour hole. The width of the jet impinging area shall not affect the stability of river banks or other structures.

5.4.3 If the hydraulic jump stilling basin is adopted for energy dissipation, stable jump shall be ensured to avoid the occurrence of backflow. The stilling basin and the areas upstream and downstream of the end sill shall be cleaned of accumulated debris.

5.4.4 The stilling basin is desired to have a rectangular cross-section with equal width. When the topography allows, a sloping apron may be adopted. When the average flow velocity before the hydraulic jump is less than 16 m/s, baffle piers and/or other appurtenances may be provided on the apron. In the cold region, the appurtenances shall meet the requirements in the *Design Specifications for Frost Resisting of Hydraulic Structures* (SL 211).

5.4.5 The crest elevation of the sidewalls of the stilling basin may be determined based on the water depth after the jump plus a freeboard. If there is a certain depth of water on the outside of the sidewalls, the wall height may be appropriately decreased to permit water with small height overtopping the walls.

5.4.6 The flow regimes of energy dissipation by means of "surface-type energy dissipator" and roller bucket are complicated and unstable. The following measures are desired to prevent the dam foundation from undermining and the downstream banks from scouring for ensuring the safety of the project:

   a A key wall beneath the bucket lip or a short apron downstream of the bucket lip.

b Training walls with certain length to prevent transverse backflow.

c Bank protection downstream.

d Uniform opening of gates to discharge flow.

5.4.7 When two or more types of energy dissipators are employed jointly, the scouring resistant devices may be designed according to the provisions of Clause 5.4.1 to Clause 5.4.6. In the case of a combination of flaring pier and hydraulic jump stilling basin, hydraulic model tests shall be performed to determine the pulsating pressure distribution on the apron slab of the stilling basin, and the test results may be based for strengthening the apron slab and enhancing the integrity of the slab itself and its bonding to the foundation. Besides, measures shall be taken to ensure the reliability of the waterstops installed in the apron slab.

5.4.8 In the selection of energy dissipators, the impact of heavy spray on the operation safety of other structures and on the stability of bank slopes shall be studied, especially for the dam constructed in a narrow valley or in arid region. The buildings, outdoor electrical equipment, transmission lines, access road, etc. at the downstream area of the dam should be kept away from the intensive spray area, unless they are effectively protected.

6 Design of Dam Cross Section

6.1 Loads and Loading Combinations

6.1.1 The loads acting on the dam are grouped into basic loads and special loads:

    a Basic loads

    1) Dead loads—the weight of the dam body and permanent equipment on the dam.

    2) Hydrostatic pressures on the upstream and downstream faces of the dam at normal pool level or design flood level (selecting one of them as critical).

    3) Uplift pressure.

    4) Silt pressure.

    5) Wave pressure at normal pool level or design flood level.

    6) Ice pressure.

    7) Earth pressure.

    8) Hydrodynamic pressure at design flood level.

    9) Other frequent loads.

    b Special loads

    10) Hydrostatic pressures on the upstream and downstream faces of the dam at maximum design flood level.

    11) Uplift pressure at maximum design flood level.

    12) Wave pressure at maximum design flood level.

    13) Hydrodynamic pressure at maximum design flood level.

    14) Earthquake loads.

    15) Other infrequent loads.

The formulae for calculation of these loads are given in Annex B.

6.1.2 The loading combinations used for the analysis of dam stability against sliding and for the calculation of dam stresses are categorized into two cases, the usual cases and unusual cases. Loading combinations shall be considered as per those tabulated in Table 6.1.2. Other adverse loading combinations shall also be considered if necessary.

Table 6.1.2 Loading combinations

Loading combinations	Load case	Loads										Remarks
		Dead load	Hydrostatic pressure	Uplift pressure	Silt pressure	Wave pressure	Ice pressure	Earthquake load	Hydrodynamic pressure	Earth pressure	Other loads	
Usual	(1) Normal pool level	1)	2)	3)	4)	5)	—	—	—	7)	9)	The earth pressure, depending on whether there is earthfill against the dam face (the same below).
	(2) Design flood level	1)	2)	3)	4)	5)	—	—	8)	7)	9)	The hydrostatic and uplift pressures shall be calculated with the corresponding reservoir water level in winter.
	(3) Freezing	1)	2)	3)	4)	—	6)	—	—	7)	9)	
Unusual	(1) Maximum design flood level	1)	10)	11)	4)	12)	—	—	13)	7)	15)	
	(2) Earthquake	1)	2)	3)	4)	5)	—	14)	—	7)	15)	The hydrostatic pressure, uplift pressure, and wave pressure shall be calculated with the normal pool level, which may be other water level if justified.

Notes: 1. The most unfavorable loading combination shall be selected based on practicable possibility of simultaneous occurrence of various loads.
2. Dams to be constructed in stages shall have their corresponding loading combination calculated for each stage.
3. Loading combination during construction period shall be examined as an unusual load case.
4. If, according to geology or other conditions, the drainage system in the dam is susceptible to blocking and needs to be repaired during operation, loading combination with drains inoperative shall be considered, as an unusual case.
5. For earthquake condition, if ice pressure in winter is considered, wave pressure shall be excluded.

6.2 Design Principles

6.2.1 The cross section of a concrete gravity dam shall be determined based on the gravity method and the rigid-body limit equilibrium method, and also by the finite element method as a supplementary method.

The stresses on the upstream and downstream faces of a solid gravity dam are calculated by the equations using gravity method as presented in Annex C.

For high dams or medium dams built on complicated foundations, finite element method analysis should be carried out. If necessary, physical model tests may be carried out for verification.

The stresses around complex structures, such as openings in dam, should be calculated using finite element method to determine the reinforcement required.

6.2.2 The design cross section of a gravity dam shall be determined dominantly with usual loading combinations, and checked with unusual loading combinations. When the unusual loading combination is used for checking, the spatial action of the dam or other appropriate measures may be taken into account. The unusual loading combination should not be used to dominate the design cross section.

6.2.3 The stresses in a slotted gravity dam may be analyzed by gravity method. And the stresses in the upstream dam head may also be computed by finite element method. Tensile stress is allowed to occur in localized zone far from the upstream face, but it shall not exceed the allowable tensile stress of dam concrete.

6.2.4 The stresses in a hollow gravity dam may be calculated with structural mechanics, gravity method or finite element method. The configuration of the hollow gravity dam shall be optimized to the greatest possible extent to avoid the occurrence of adverse stress distributions.

6.2.5 A gravity dam provided with transverse joints should be considered as a two-dimensional structure in the strength and stability analyses. One dam monolith or a unit length of dam segment may be taken for the analyses.

A gravity dam with grouted transverse joints should be treated as a three-dimensional structure and the design may consider the overall behavior of the entire dam.

6.3 Stress Computation

6.3.1 Stresses to be computed mainly include:

  a Stresses in the selected horizontal planes (these planes shall be selected based on the dam height, and the selected planes shall include the dam-foundation interface, the planes at which the dam slope changes, and any other planes that need to be analyzed).

  b Localized stresses in the weakened parts of the dam (such as openings or orifices, conduits, and power waterway).

  c Stresses at the individual locations of the dam (such as gate piers, breast walls, training walls, supporting structures of the intake, and upstream head of the slotted gravity dam).

  d Stresses within the dam foundation, if required.

  e Stresses which may include some or all of the above-mentioned items, or any other added item, depending on the project size and the dam structure.

6.3.2 The vertical stresses on the dam-foundation interface at the dam heel and dam toe shall meet the following requirements:

  a Operation period:

  1) **Under all loading combinations (exclusive of earthquake loads), the vertical stresses at the dam heel shall not be tensile stresses, and the vertical stresses at the dam toe shall be less than the allowable compressive stresses of the dam foundation.**

  2) **Under the earthquake loads, the vertical stresses at the dam heel and dam toe shall meet the relevant requirements in *Specifications for Seismic Design of Hydraulic Structures* (SL 203).**

  b Construction period:

  The vertical tensile stress of less than 0.1 MPa is allowable at the dam toe.

6.3.3 The vertical stresses on the dam-foundation interface shall be calculated according to Equation (6.3.3):

$$\sigma_y = \frac{\Sigma W}{A} \pm \frac{\Sigma Mx}{J} \quad (6.3.3)$$

where σ_y—vertical stresses on the dam-foundation interface, kPa;

ΣW—summation of all normal forces (including uplift, the same below) acting on the dam-foundation interface of one dam monolith or one unit length of dam monolith, kN;

ΣM—summation of moments of all forces about the centroidal axis of the dam-foundation interface under consideration, kN · m;

A—area of the dam-foundation interface of one dam monolith or one unit length of dam monolith, m²;

x—horizontal distance from the centroidal axis of the interface to the point under consideration, m;

J—moment of inertia of the interface of one dam monolith or one unit length of dam monolith about its centroidal axis, m⁴.

6.3.4 The stresses of a gravity dam shall meet the following requirements:

a Operation period:

1) **The vertical stresses at the upstream face shall not be tensile stresses (with the uplift pressure considered).**

2) **The major principal compressive stress shall not exceed the allowable compressive stress of the concrete.**

3) **In the case of earthquakes, the stresses at the upstream face shall meet the relevant requirements in** *Specifications for Seismic Design of Hydraulic Structures* **(SL 203).**

4) **Provisions on the tensile stresses in local areas of the dam:**

i For a slotted gravity dam, tensile stresses are allowed to occur in local areas far from the upstream face, but they shall not exceed the allowable tensile stress of the concrete.

ii If the tensile stresses occur at the weir crest portion of the overflow dam, that particular portion shall be strengthened by steel bar reinforcement.

iii The tensile stress zones around the galleries and other openings should be strengthened with steel bar reinforcement. Nevertheless, less or no reinforcement may be needed if it is justified by careful study.

b Construction period:

1) **The principal compressive stresses in any section of the dam shall not exceed the allowable compressive stress of the dam concrete.**

2) **The principal tensile stress not greater than 0.2 MPa is permitted at the downstream face of the dam.**

6.3.5 The influence of longitudinal joints may be neglected in calculating the stresses in the gravity dam. However, for high dams or dams with an overhanging upstream face, the stress condition in the upstream construction block before grouting of the longitudinal joint shall be considered with care, and measures shall be taken to restrict and improve the adverse stress condition.

6.3.6 For dam monoliths on the abutments, the stability against sliding and the stresses under the combined action of loads from three directions shall be analyzed by taking account of the topography, geology, and foundation excavation. In such case, small tensile vertical stresses are allowed to occur at the dam heel if it is unavoidable. However, measures shall then be adopted to make sure that the dam satisfies the requirements in sliding stability and stress condition during operation and construction periods.

6.3.7 Finite element method may be used for the stress analysis of dam monoliths or sections that cannot be simplified as a two-dimensional problem, or for the stress analysis of any other structures that cannot be analyzed with the gravity method. In this case, the values of local stresses may not be limited by those prescribed in Clause 6.3.4. However, if the local stresses are overly high, the causes for the high values shall be identified and proper measures shall be taken to improve the stress conditions.

6.3.8 When finite element method is used to calculate the dam stresses, the finite element mesh shall contain enough elements to meet the required accuracy of the design. The shapes of the elements shall be rationally selected to fit the shape of the dam. The computational models and conditions shall have a close simulation of the actual conditions.

6.3.9 When stresses at the dam base are computed by the finite element method, the width of tensile stress zone near the upstream face should be smaller than 0.07 times the base width, or smaller than the distance from the centerline of grout curtain to the dam heel.

6.3.10 The allowable stress of the concrete shall be determined by dividing the ultimate strength by a corresponding safety factor.

The safety factor for determining the allowable concrete compressive stresses shall not be smaller than 4.0 for usual loading combination and 3.5 for unusual loading combination (exclusive of earthquake load).

When the concrete in local areas is required to resist tensile stresses, the safety factor for determining the allowable concrete tensile stresses shall not be less than 4.0.

In the case of earthquakes, the safety of the dam structure shall meet the requirements in *Specifications for Seismic Design of Hydraulic Structures* **(SL 203).**

Note 1: The ultimate compressive strength of concrete refers to the strength of 15-cm cubes at 9-day age, with an 80% probability of assurance.

Note 2: The design and calculation of local structure in the dam shall follow the requirements specified in *Design Code for Hydraulic Concrete Structures* (SL/T 191).

6.4 Calculation of Dam Stability against Sliding

6.4.1 Calculation of the dam stability against sliding shall be performed mainly for the dam base (dam-foundation interface). The safety factors of stability against sliding at the dam-foundation interface shall be calculated by either the shear-friction Equation (6.4.1-1) or the friction Equation (6.4.1-2).

a Safety factor using shear-friction equation:

$$K' = \frac{f' \Sigma W + C'A}{\Sigma P} \quad (6.4.1\text{-}1)$$

where K'—factor of safety of stability against sliding based on shear-friction equation;

f' —friction coefficient at the dam-foundation interface based on shear-friction equation;

C' —unit cohesion at the dam-foundation interface based on shear-friction equation, kPa;

A —area of the interface, m^2;

ΣW —summation of normal components of all forces on the sliding plane, kN;

ΣP —summation of tangential components of all forces on the sliding plane, kN.

b Safety factor using friction equation:

$$K = \frac{f \Sigma W}{\Sigma P} \quad (6.4.1\text{-}2)$$

where K —factor of safety of stability against sliding based on friction equation;

f —friction coefficient at the interface based on friction equation.

c Stipulations for the safety factors of stability against sliding:

1) The factors of safety K' determined by Equation (6.4.1-1) shall not be less than the values specified in Table 6.4.1-1.

Table 6.4.1-1 Safety factor K' of stability against sliding at the interface

Loading combination		K'
Usual loading combination		3.0
Unusual loading combination	(1)	2.5
	(2)	2.3

2) The factors of safety K determined by Equation (6.4.1-2) shall not be less than the values specified in Table 6.4.1-2.

d When weak structural planes or gently dipping fissures exist in the foundation rock, the factor of safety of stability against sliding along such deep-seated planes or fissures shall be calculated as specified in Annex E. The values of K' determined by Shear-Friction Equation (E.0.2-1) and Equation (E.0.2-2) shall not be less than the specified values in Table 6.4.1-1. If the value of K' still cannot meet the requirements specified in Table 6.4.1-1 even after engineering measures are taken for foundation treatment, the factor of safety of stability against sliding along deep-seated planes may be calculated by Friction Equation (E.0.3-1) and Equation (E.0.3-2). The allowable factor of safety for such case may be determined only after careful justification. In this context, special caution shall be paid to the dam stability against sliding along a single sliding plane.

Table 6.4.1-2 Safety factor K of stability against sliding at the interface

Loading combination		Class of dam		
		1	2	3
Usual loading combination		1.10	1.05	1.05
Unusual loading combination	(1)	1.05	1.00	1.00
	(2)	1.00	1.00	1.00

6.4.2 The friction coefficient f' and cohesion C' and the friction coefficient f on the dam-foundation interface may be selected by referring to Annex D in the project planning phase, but they shall be determined by tests in the feasibility study phase and subsequent phases. In the case of a medium or low dam for a medium-sized project, when the in-situ tests are not possible to be conducted, these coefficients should be determined through laboratory tests along with reference to Annex D.

6.4.3 In the stability analysis of dam, the combined action of the dam and the powerhouse or other massive concrete structures located at the dam toe may be considered after due justification. In such a case, corresponding structural design shall be performed accordingly.

6.4.4 When a weak structural plane or a gently dipping joint exists in the foundation rock, the stability of the dam against sliding along these deep-seated planes shall be checked. According to the distribution of slip planes, the computational models may be classified into the single-plane, double-plane, and multiple-plane sliding modes. The analysis shall be done primarily with rigid body limit equilibrium method (refer to Annex E). If necessary, the finite element method and geomechanical model tests may be adopted to analyze and evaluate the sliding stability along weak planes in the foundation. The findings of these analyses may form a basis for selecting the dam foundation treatment program.

When no unfavorable faults or fissures parallel to the stream direction exists in the dam foundation, and the transverse joints in the dam are keyed and grouted, the sliding resistance from adjacent dam monoliths may be considered for checking the stability of sliding along deep-seated weak planes in the rock foundation.

6.5 Structural Design of Gate Piers on Overflow Dam Section

6.5.1 The strength calculation of the gate piers on the overflow dam section includes:

a Checking the longitudinal strength of the gate pier, when the pier is bearing the maximum longitudinal forces, with the corresponding lateral forces, vertical forces and dead weight.

b Checking the transverse strength of the gate pier, when the pier is bearing the maximum unbalanced lateral forces, with the corresponding longitudinal forces, vertical forces and dead weight.

c Checking the strengths of the portion where the gate slots and the trunnion of the radial gate are placed.

d Checking the displacement of the gate pier, if necessary.

6.5.2 The strength calculation of the gate piers shall meet the following requirements:

a In checking the longitudinal strength, the tensile stress shall not occur in the pier and reinforcing bars at the perimeters of the pier may be provided according to the structural requirements or other related conditions. If the tensile stresses are hardly avoidable, the pier shall be designed as a reinforced concrete structural element under slight eccentric compression.

b In checking the transverse strength, the pier shall be regarded as an integral element with a fixed end, and designed as a reinforced concrete structural element under eccentric compression or eccentric tension.

c Crack control in the localized tensile zone around the trunnions of a radial gate and the shear-span ratio of the cross-section of the trunnions shall meet the structural requirements in design.

d In the case of earthquake load, the gate pier shall meet the strength requirements in *Specifications for Seismic Design of Hydraulic Structures* (SL 203).

6.5.3 Prestressed concrete structure may be adopted when the gate pier supporting the radial gate is subject to relatively large force.

7 Foundation Treatment

7.1 General

7.1.1 The foundation of a concrete gravity dam after proper treatment shall satisfy the following requirements:

a Have adequate strength to bear the pressure of the dam body.

b Have adequate integrity and uniformity to meet the requirements of dam stability against sliding and to decrease the differential settlement.

c Have adequate imperviousness to satisfy the seepage control requirement, to limit the seepage flow and reduce the seepage pressure.

d Have adequate durability to prevent deterioration of rock properties under the long-term water action.

7.1.2 In the design of dam foundation treatment, the interaction between the foundation and the superstructure shall be taken into comprehensive consideration. If required, measures may be taken to adjust the type of the superstructure so as to achieve compatibility in working conditions of the superstructure and the foundation.

7.1.3 In the design of dam foundation treatment, study of the slope stability, the deformation and seepage should be made for the abutments and nearby upstream and downstream areas. If necessary, corresponding treatment measures shall be adopted according to the study.

7.1.4 In Karst area, special foundation treatment shall be designed after the Karst caverns and wide solution fissures in the foundation are carefully investigated with respect to their locations, formation characteristics, filler properties and groundwater movement.

7.2 Foundation Excavation

7.2.1 The position of the dam foundation surface shall be determined through technical and economic comparison based on such factors as dam stability, foundation stresses, physical and mechanical properties of rock mass, classes of rock, foundation deformation and stability, requirements of the structure on the foundation, effectiveness of foundation treatment, construction technology, construction period, costs and so on. Measures may be taken to reduce the volume of foundation excavation by strengthening the foundation and/or modifying the structure on the condition that the foundation strength and stability is not compromised.

A dam over 100 m in height may be founded on fresh bedrock, or on the lower part of slightly to moderately weathered bedrock; a dam between 50 m and 100 m in height may be founded on the middle part of slightly to moderately weathered bedrock; a dam less than 50 m in height may be founded on the middle to upper part of weakly weathered bedrock. These requirements may be appropriately less stringent for the dam sections resting on higher parts of the abutments.

7.2.2 The foundation surface of a concrete gravity dam shall be shaped in accordance with the site topographical and geological conditions, and also the superstructure requirements. The foundation surface of a dam monolith should not have great difference in height at its upstream and downstream sides and is desired to be slightly inclined toward upstream. If the foundation surface has great difference in height, or inclines toward downstream, it should be excavated into large steps with blunt corners. The height difference between the steps shall be consistent with the sizes of concrete placement blocks and the locations of construction joints; it shall also be adaptable to the concrete thickness at the dam toe. Where there is excessive height difference at the foundation portion, the joints between the dam monoliths should be adjusted, or necessary treatment may be adopted.

7.2.3 The foundation base of dam monoliths on the abutments shall be excavated into terraces with adequate width in the dam axis direction, or other structural measures shall be adopted to ensure the lateral stability of the dam.

7.2.4 Localized geological defects in the foundation, such as superficial mud-filled fissures, weathered pockets, fault-fractured zones, closely jointed zones, Karst fillings, and shallow-seated weak planes, shall be totally removed during foundation excavation, or partly removed and then treated properly.

7.2.5 In foundation excavation design, criteria for blasting operations shall be proposed to ensure that the foundation rock will not be damaged or adversely affected. Protective measures shall be provided for the rock masses that are easily susceptible to weathering or becoming argillized upon exposure.

7.3 Consolidation Grouting of Dam Foundation

7.3.1 Consolidation grouting in the dam foundation shall be designed on the basis of geological conditions of the foundation, the height of the dam, and the results of grouting tests.

a Consolidation grouting should be carried out in a certain range at upstream and downstream areas inside of the dam foundation. Consolidation grouting may cover the entire foundation area, if joints or fissures are well developed and are groutable in the foundation rock mass. Consolidation grouting may be extended appropriately to outside the dam base, and to the slot areas in the case of a slotted gravity dam depending on the foundation stresses and geological conditions.

b Consolidation grouting should be applied to the foundation area upstream of the grouting curtain.

c Fault fracture zones and their influence areas, and other geological defects shall be treated with intensified consolidation grouting.

d The Karst caverns and trenches in the dam foundation shall be treated by removing filling and backfilling the voids, and then the surrounding areas shall be strengthened by consolidation grouting in accordance with the Karst distribution.

7.3.2 The spacing of the grout holes and the spacing of the hole lines parallel to the dam axis may vary from 3 m to 4 m, or they may be determined by grouting tests to be conducted after the geological conditions are exposed by excavation.

The depths of grout holes shall be determined based on the dam height and on the geological conditions in the foundation exposed by excavation. They are generally varying from 5 m to 8 m, and may be increased, if necessary, for some local areas or for large foundation stresses of high dam foundations. The depths of grout holes upstream of the grout curtain are normally taken between 8 m and 15 m, depending on the curtain depths.

7.3.3 The grout holes are usually arranged in a staggered manner. Special grout holes shall be provided in large faults and fractured zones. The orientation and inclination of grout holes shall be determined based on the main features of the joints and fissures in combination with the construction condition for letting the holes intersect as many joints and fissures as possible.

7.3.4 For foundation upstream of the grouting curtain or with geological defects, it is preferred that consolidation grouting be performed with 3 m to 4 m thick counterweight concrete. For other parts, consolidation grouting may also be performed on top of counterweight concrete, or just on foundation rock surface, or on leveling concrete only, after due justification.

7.3.5 Under the condition of no uplifting the foundation rock and counterweight concrete, the consolidation grouting pressure should be as high as practicably possible. When grouting on the counterweight concrete, the pressure may be taken between 0.4 MPa and 0.7 MPa, depending on the concrete thickness. When grouting is conducted on concrete-leveled and sealed foundation surface, the pressure will normally be determined by grouting tests, generally ranging from 0.2 MPa to 0.4 MPa. For the foundation composed of rocks with gently dipping structural planes or weak rocks, the grouting pressure shall be determined by grouting tests.

7.4 Seepage Control and Drainage of Foundation

7.4.1 Seepage control and drainage of dam foundation shall be designed on the basis of engineering geological conditions, hydro-geological conditions, and grouting test results. Corresponding measures for seepage control and for drainage shall be determined with comprehensive consideration of the adaptability and combined action of these two measures in conjunction with the reservoir functions and the dam height. For high dams with complicated hydro-geological conditions, seepage analysis shall be conducted.

7.4.2 Cement grouting may be used as a measure for seepage control of dam foundation and abutments. Cement-mixed materials grouting, as well as chemical materials grouting (if necessary) may also be employed. If applicable by verification, concrete cutoff wall, or concrete diaphragm wall, or horizontal impervious blanket may also be adopted for seepage control of dam foundation. Cutoff wall formed by concrete backfilling of open excavation or shaft excavation may also be used for the seepage control of abutments. On a silt-laden river, if the deposited sediment in the reservoir is impervious and thick enough, its positive effect on reducing seepage may be taken into account in the design of seepage control in an appropriate way; however, the safety of dam during initial operation period shall be guaranteed.

7.4.3 The grout curtain for seepage control shall meet the following requirements:

a Reduce seepage through dam foundation and around abutments to avoid the adverse effect of seepage flow on the foundation and abutments.

b Prevent seepage failure occurring in weak structural planes, fault-fractured zones, fissures, or rock mass with poor seepage erosion resistance.

c Reduce the uplift pressure and seepage flow to the allowable range under the combined action of grout curtain and drainage.

d Maintain the continuity and adequate durability.

7.4.4 For a large and medium project, or a high dam, curtain grouting tests shall be performed before commencement of the curtain grouting. The grout curtain design may be modified based on the borehole data gathered during construction. The primary grout curtain shall be completed before reservoir impoundment.

7.4.5 The permeability (q, expressed in Lugeon/Lu) of the grout curtain and the relatively impervious rock strata shall meet the following criteria:

a For dams over 100 m high, q is between 1 Lu and 3 Lu.

b For dams between 100 m and 50 m high, q is between 3 Lu and 5 Lu.

c For dams below 50 m high, q is 5 Lu.

d For pumped-storage power stations or reservoirs in areas of shortage in water resources, the lower limit of the above ranges shall be adopted.

7.4.6 The depths of grout curtain shall comply with the following provisions:

a Closed grout curtain: when a reliable impervious stratum is not deeply seated in the foundation, the grout curtainshall extend 3-5 m into the impervious stratum. The permeability values of the relatively impervious strata for dams of different height are specified in Clause 7.4.5.

b Hanging grout curtain: when the relatively impervious stratum is deeply seated or irregularly distributed, the curtain depths shall satisfy the requirements of Clause 7.4.3, and shall be determined based on comprehensive considerations of seepage analysis results, engineering experiences, geological conditions, uplift pressure on the dam base and other factors. Normally, the curtain depths may be selected in a range from 0.3 to 0.7 times the water head.

7.4.7 When the grout curtain is rather deep at the dam abutments and banks, grouting tunnels shall be drilled at different levels. The layout of the tunnels shall be determined by considering the topographical and geological conditions, drilling and grouting technique, and ventilation and drainage in construction, as well as the elevation distribution of Karst caverns if applicable. The difference in elevation between the grouting tunnels may range from 30 m to 60 m. The curtains grouted from different tunnels may be connected in an inclined, straight, or staggered manner to ensure the connection being continuous, closed and tight.

7.4.8 The length that the curtain shall extend into the abutments and the orientation of the curtain shall be determined on the basis of geological and hydro-geological conditions. The curtain should extend to the relatively impervious strata or to the intersection of normal pool level with groundwater level. The curtain in the dam abutments shall maintain a continuous connection with that in the riverbed.

7.4.9 The number of lines of grout holes, the spacing of lines and that of grout holes shall be determined in light of geological and hydro-geological conditions, water head, and grouting test results.

When theactive role in seepage reduction in the shallow part of the foundation upstream of the grout curtain by consolidation grouting is taken into consideration, a double-line grout curtain may be adopted for dams with a height of 100 m or higher, and a single-line curtain may be adopted for dams with a height of less than 100 m. The lines of curtain may be suitably increased in areas with very poor geological conditions and highly developed fissures, or in areas with probable seepage erosion, or in areas where the study considers necessary to enhance the curtain.

When a multiple-line curtain is adopted, one of the lines of the grout holes shall be drilled and grouted to the design depth, and the other lines may be drilled and grouted to 1/2-2/3 of the design depth.

The spacing of the curtain grout holes may range from 1.5 m to 3 m and the spacing of curtain lines should be slightly less than the spacing of grout holes.

The grout holesare desired to intersect the main fissures and bedding planes of the rock masses and may be inclined 0°-10° toward upstream from the vertical.

7.4.10 Curtain grouting must be performed only after a certain thickness of dam concrete has been placed to serve as counterweight. The grouting pressure shall be determined by grouting tests. As a general rule, the grouting pressure is taken to be 1.0-1.5 times the reservoir head for the top stage, and may be gradually increased for the lower stages, up to 2.0-3.0 times the reservoir head for the bottom stage. Anyhow, the dam concrete or foundation rock shall not be uplifted during grouting operation.

7.4.11 The main drain holes in the dam foundation are generally arranged in the grouting gallery downstream of the grout curtain. It is not desired that the distance from the main drain holes to the grout curtain holes measured at the dam base elevation be less than 2 m. For high dams, 2 or 3 lines of drain holes may be drilled as auxiliary drainage, and for medium dams, 1 or 2 lines will do. If necessary, the drain holes may be provided along the transverse drainage gallery or at the slots (in the case of a slotted gravity dam). When relatively impervious strata and gently dipping bedded planes exist in the foundation, their distribution shall be considered for making rational arrangement of the drain holes.

7.4.12 For dams with high tailwater level and provided with a pumped drainage system, auxiliary drain holes both in longitudinal and transverse direction shall be arranged in the foundation downstream of the main drain holes. When the high tailwater level maintains for a long duration or the downstream foundation rocks are fairly permeable, a laterally enclosed grout curtain is preferred at the dam toe.

7.4.13 For low dams founded on sound and weakly pervious rock foundation (with the coefficient of permeability less than 0.1 m/d), the grout curtain may not be provided, and only drain holes are arranged to reduce the seepage pressure. However, consolidation grouting shall be conducted in the upstream area of the dam foundation.

7.4.14 The spacing may vary from 2 m to 3 m for main drain holes and 3 m to 5 m for auxiliary drain holes.

7.4.15 The depths of the drain holes shall be determined on the basis of the depths of grout curtain and consolidation grouting, and the geological and hydro-geological conditions of the foundation.

a The depth of the main drain holes shall be 0.4-0.6 times the curtain depth. For high and medium dams, the main drain holes shall not be less than 10 m deep. When fissured aquifer or deep pervious stratum exists in the dam foundation, the main drain holes should penetrate into this stratum, in addition to taking measures to strengthen the seepage control.

b The depths of auxiliary drain holes may be 6-12 m.

7.4.16 The foundation of abutment dam sections may be provided with special drainage system. If necessary, drainage tunnels may be driven into the abutments for providing the drainage holes.

7.4.17 When the drain holes are subject to wall sloughing, or when they intersect the weak structural planes and mud-filled fissures, protection measures shall be correspondingly taken for such holes.

7.5 Treatment of Fault Fracture Zones and Weak Structural Planes

7.5.1 The fault fracture zones or weak structural planes in the foundation shall be specially treated according to their position, depth, attitude, width, composition properties, laboratory test results, and their influence on the superstructure.

In earthquake regions with an intensity of Ⅷ or higher, the treatment requirements shall be more stringent.

In the case of a low dam, the requirements for fracture zone treatment may be less stringent.

7.5.2 Steeply dipping fault fracture zones may be treated with the following methods:

a When an individual outcrop of fault fractured zone with main composition of hard tectonite exists within the dam foundation and when it will not considerably impact the strength and compression deformation of the foundation, the fault fractured zone and the affected zones along the both sides may be removed by excavation to a proper extent.

b When the fault fractured zone is of a small scale, but the main composition is weak tectonite, which may cause certain impact on the foundation strength and compression deformation, such zones may be reinforced by the method of dental treatment. The depth of the dental treatment may be taken as 1.0-1.5 times the width of the fault fractured zone, or it may be determined through calculation. When the fault fractured zone stretches through the dam foundation from upstream to downstream in the river flow direction, such treatment should extend to areas beyond the dam base on both upstream and downstream sides.

c When either a fault fractured zone or an intersection fault zone is of a large-scale with the main composition of weak tectonite, and may have great effect on the foundation strength and compression deformation to a large extent, special design for treatment of such zone must be prepared.

7.5.3 The stability against sliding along deeply-seated weak planes may be improved on principle by the following methods:

a Increase the shearing resistance of the weak structural planes.

b Increase the resistance of resisting rock wedges.

c Combine the above two measures.

d Select the treatment measures after comprehensive analysis and comparison on the attitude, depth, characteristics of the weak structural planes, and their impact on the dam, in combination with the project size, construction conditions and progress.

7.5.4 The shearing resistance capacity of weak structural planes may be improved by concrete replacement, or deep concrete cut-off wall, or concrete-plugged tunnels or shaft, to be selected according to the embedded depths of the planes. If necessary, shear piles, prestressed tendons, chemical grouting, etc. ,may also be adopted.

7.5.5 When a large concrete plug, a concrete cutoff wall, or a concrete-plugged tunnel needs to use large amount of concrete for treating the gentle-dipping weak structural plane, temperature control measures shall be adopted and contact grouting shall also be performed.

7.5.6 If the fault fractured zone or weak structural plane extends into the reservoir area and may create a leakage passage and worsen the geological conditions, special treatment shall be provided to control the seepage.

7.5.7 The foundation drainage facilities at the fault fractured zones or weak structural planes shall be determined on the basis of the geological conditions and shall also conform to the provision of Clause 7.4.17.

7.6 Seepage Control in Karst Areas

7.6.1 The methods for seepage control in Karst areas include curtain grouting, diaphragm wall (or cut-off wall) and others. They shall be selected on the basis of the scale, distribution, permeability of the Karst formation, and the properties of fillings. Karst caverns or highly permeable solution fissures may be treated by excavation and concrete backfilling, or by high pressure grouting after provision of grout seal holes (wells), or other measures.

7.6.2 If a Karst cavern crossing the dam foundation from upstream to downstream is embedded at shallow depth or wherever feasible in construction condition, such cavern shall be treated by excavation and concrete backfilling. If such a Karst cavern is deeply seated and is difficult to be removed by open excavation, it may be treated by tunnel excavation and concrete backfilling, and also by trench excavation and concrete backfilling.

7.6.3 The alignment of grout curtain at the abutments shall be determined on the basis of the topographical and geological conditions, and the Karst development characteristics. In plan, it may be a straight line, a broken line, or a line with forward or backward wings. In geologically complicated areas, the grout curtain alignment shall be determined through technical and economic comparison of different schemes, and if necessary, the selection may be incorporated with the comparison of the dam axis. Efforts shall be made to align the grout curtain traversing the areas with less karst formations. If the grout curtain line has to cross the Karst underground river or Karst passage, it should intersect them perpendicularly as long as it is possible.

7.6.4 In Karst areas, the depth of grout curtain shall be determined through technical and economic comparison, by considering such factors as the depth of relatively impervious stratum, dam height, allowable leakage through dam foundation and abutments, uplift pressure downstream of the grout curtain, and other factors. Anyhow, dam safety shall be ensured.

7.6.5 The number of lines of a grout curtain, the spacing of lines, the spacing of grout holes and the grouting pressure shall be determined on the basis of geological structure and hydro-geological conditions of the Karst formation in conjunction with grouting tests. The grouting tests shall include the studies on the allowable seepage gradient and durability of the curtains formed in different types of Karst caverns and different filling materials in the caverns.

7.6.6 The grouting materials may be selected according to the scale of Karst caverns and solution fissures, as well as the fillings. They include neat cement grout, cement-sand grout, cement-clay grout, cement-fly ash grout and others. If necessary, large-diameter boreholes may be drilled and then grouted by using the fine-aggregate concrete with high fluidity.

8 Structural Arrangement of Dam Body

8.1 Dam Crest

8.1.1 The crest level of a dam shall be higher than the maximum flood level, and the top of the upstream wave wall or parapet on the dam crest shall be higher than the peak level of the wave. The height differences between the top of wave wall and the normal pool level or the maximum flood level may be calculated by Equation (8.1.1). The higher value of the two wave wall top elevations shall be selected as the top elevation of the wave wall.

$$\Delta h = h_{1\%} + h_z + h_c \qquad (8.1.1)$$

where Δh—height difference between the top of wave wall and the normal pool level or the maximum flood level, m;

$h_{1\%}$—wave height, m;

h_z—height difference between the wave centerline and the normal pool level or the maximum flood level, m;

h_c—freeboard, m, taken from Table 8.1.1.

Table 8.1.1 Freeboard, h_c

Water level	Class of dam		
	1	2	3
Normal pool level	0.7	0.5	0.4
Maximum flood level	0.5	0.4	0.3

8.1.2 The wave wall should be a reinforced concrete structure integrated with the dam body, and it shall be thick enough to resist the impact of waves and floating debris. The joints shall be provided on the wave wall at the locations of the transverse joints of the dam monolith and shall be sealed with waterstops. The height of the wave wall may be taken as 1.2 m. Handrails shall be installed along the downstream side of the dam crest.

8.1.3 The crest width of non-overflow dam sections shall be established on the basis of dam cross-section design and the operation requirements. Anyhow, the crest width should not be less than 3 m. The roadway on the dam crest shall have a transverse slope and drainage. In cold regions, the transverse slope of the roadway shall be properly increased.

8.1.4 On the crest of overflow dam sections, the service bridge and access bridge shall be provided taking into account the arrangement of the gates and hoisting equipment, and the requirements for operation and maintenance, traffic, monitoring, etc. The bridge may be built as a prefabricated reinforced concrete or a prestressed reinforced concrete structure. Sufficient clearance shall be furnished beneath the bridge.

8.1.5 The service and access bridges on the overflow dam sections shall meet the requirements of *Specifications for Seismic Design of Hydraulic Structures* (SL 203).

8.1.6 When the dam crest is used as a public roadway, the footpath provided at the roadway side should be 30 cm higher than the roadway surface.

8.1.7 The layout of the dam crest structure shall be incorporated with the overall project layout and shall be in harmony with the surrounding environment.

8.2 Galleries and Access Passages in Dam

8.2.1 The galleries are provided inside the dam for various purposes, such as foundation grouting, drainage, safety monitoring, inspection and maintenance, operation work, internal access, and construction requirement.

8.2.2 Grouting galleries must be provided in a high or medium dam. The longitudinal slope of the grouting gallery shall be gentler than 45°. In the case of a relatively steep gallery with great length, intermediate safety platforms with handrails shall be provided at different intervals. If the slopes of the abutments are steeper than 45°, the grouting gallery at abutment may be arranged in numbers of horizontal subsections at different elevations and connected by vertical shafts.

Generally, the floor slab of the grouting gallery should have a thickness not less than 3 m.

8.2.3 When multi-level galleries are provided within the dam, the height difference between two levels should be in the range from 20 m to 40 m in general, and access shall be provided in between the upper and lower galleries. The distance from the gallery to the openings of outlet works in the dam should not be less than 3-5 m, and shall be determined by stress analysis to avoid cracking in concrete. The minimum distance between the upstream gallery wall and the upstream dam face shall meet the requirements for seepage control (usually 5%-10% of the reservoir head on the upstream dam face), and shall not be less than 3 m.

8.2.4 Access bridges should be provided at the downstream dam face at different elevations above the highest tailwater level. The elevations of these bridges should be consistent with the positions of openings and galleries in the dam. The entrances or exits of the galleries shall be furnished with doors for security and cold-proof purposes. Measures shall be adopted to prevent water from entering the galleries during flood releasing in both construction period and operation period. Elevators in the shafts are the major means of vertical access inside the dam. For a high dam of a large project, at least two elevators and stairways should be provided in general; while for a medium dam, elevator may be provided according to the actual requirements.

8.2.5 The galleries should be standardized in their cross-sections as long as it is possible. The standardized cross sections can be inverted U-shaped or a rectangular. However, for the galleries running along the transverse joints, the cross section may be that with a triangular roof and a flat floor.

8.2.6 The cross-sectional size of foundation grouting galleries shall be determined by the space required for working and for drilling and grouting equipment. Generally, the grouting gallery may be 2.5-3.0 m in width and 3.0-3.5 m in height. For a long foundation grouting gallery, grouting equipment rooms may be provided along its length for conducting grouting operation. The other galleries shall also have ample size to accomplish its intended functions and to allow easy access. The minimum size shall be 1.2 m wide and 2.2 m high.

8.2.7 In the galleries, adequate lighting and ventilation shall be provided. All electrical equipment and wire in the galleries shall have reliable insulation. In addition, an emergency lighting system should be provided within the galleries.

8.3 Joints in Dam

8.3.1 Provisions of transverse, longitudinal and inclined joints in the dam shall be in conformity with the requirements of Clause 9.3.3.

8.3.2 When the transverse joints are provided to serve as contraction joints or settlement joints, they are neither keyed nor grouted. Usually, they are sealed near the upstream face by waterstops.

8.3.3 The transverse joints should be grouted completely or partially for the following conditions:

a When the transverse joints are required to be highly impervious.

b When the dam monoliths located on the steep slope are subjected to lateral loads, the adjacent dam monoliths need to be bonded as one monolith to satisfy the stability and stress requirements.

c When the river valley is narrow, a monolithic gravity dam is deemed favorable after economic and technical comparison.

d When the seismic intensity is above VIII, or when there are other special requirements, the dam needs to be built as a monolithic structure to enhance its performance of seismic resistance.

8.3.4 The spacing of the transverse joints shall be adapted to the arrangement of the penstocks, outlet works, diversion bottom outlets, and the spillways.

For dam sections at the abutments, transverse joints are desired to be located at places where there is abrupt change or any turn in topography.

8.3.5 The longitudinal joints shall be provided according to the size of dam monoliths and temperature control criteria. They may be vertical. Their joint surfaces on both sides shall be keyed and the grouting pipe systems shall be embedded for grouting to be carried out in later stage. The longitudinal joints may be terminated at a certain elevation, and if they are to be extended to the dam face, they shall perpendicularly intersect the dam face.

8.3.6 The longitudinal (or transverse) joint surface shall be divided by metal grout seals into several zones for grouting. Usually, each grout zone may cover an area of about 200-400 m^2 and have a height about 10-15 m.

8.3.7 During grouting of the longitudinal joints, the temperature of the dam body shall meet the provisions of Clause 9.3.10. After grouting, reservoir impoundment shall not start until the hardened cement grout achieves the anticipated strength.

8.3.8 The inlets and outlets of the embedded grouting pipe system (including air vent pipes) for a grout zone of the longitudinal (or transverse) joint shall be led to the galleries or near a platform. The grouting pressure for the longitudinal or transverse joint shall be determined on the basis of stress and strain conditions, usually between 0.1 MPa and 0.3 MPa at the top part. When several longitudinal joints exist in one dam monolith, they should be grouted uniformly in progress, or the downstream longitudinal joint may be grouted first. The transverse joints, if required, should also be grouted uniformly in progress.

8.3.9 The horizontal construction joints of adjacent placement blocks or lifts in one dam monolith shall be staggered. Where a horizontal construction joint will intersect the arched roof of a gallery, it may be connected to the arch base with a slope of 1 : 1-1 : 1.5. The horizontal joints above the gallery shall be located at least 1.5 m higher than the top of the gallery.

8.3.10 Inclined joints are applicable to medium and low dams, and they may be left ungrouted. When they are to be used in high dams, the applicability shall be verified.

8.4 Waterstop and Drainage

8.4.1 Waterstops shall be installed in the transverse joints (including joints of the wave wall) at locations near the upstream dam face, near the overflow surface, near the downstream dam face below the maximum tailwater level, and around the galleries or openings that cross through the joints.

8.4.2 The waterstops near the overflow surface shall be welded with the embedded metalwork at gate sill. The waterstops installed in the wave wall shall be connected to those in the dam transverse joints.

8.4.3 In the case of a high dam, a double-line waterstops shall be installed in the transverse joints near the upstream face. One drain well or other justifiable devices should be arranged between the two lines of waterstops. In the transverse joint, asphalt felt may be filled from the first line of waterstops to the upstream face. If specially required, grouting may be performed in the joint between the drain well and the second line of waterstops as supplementary measure for cutting seepage.

The waterstops in transverse joints of medium and low dams may be simplified to an appropriate extent.

8.4.4 Both the double line waterstops for the transverse joints of high dam shall be 1.0-mm to 1.6-mm thick copper sheets. The first line waterstops for the transverse joints of medium dam shall be made of copper sheet. The copper waterstops is preferred to be fabricated into the M-type shape in cross section, with each of both wings embedded in concrete for at least 20-cm to 25-cm long.

Depending on the water head, climate conditions, location to be installed, and easiness in construction, polyvinyl chloride waterstops or rubber waterstops of standard types may be used in the transverse joints near the upstream dam face subject to lower water head, near the downstream dam face below the maximum tailwater level, and around the galleries. Measures shall be taken to prevent the PVC waterstops or the rubber waterstops from deformation during installation.

8.4.5 The waterstops in transverse joints must be well connected to the foundation. They may be embedded 30-50 cm into the foundation. If required, the concrete dent for installing waterstops may be anchored on the foundation rocks by anchor bars.

8.4.6 When the dam monolith is located on steep abutment, the water sealing at the contact between the dam body and the foundations may be achieved by using the following methods:

a Provide concrete sill or dent on the abutment bedrock, and embed one side of the copper waterstop into it and the other side into the dam body.

b Perform contact grouting using pre-embedded grouting system or post-drilled grout holes after complete shrinkage of foundation concrete. If feasible, the grout holes for curtain grouting and consolidation grouting may be utilized to perform the contact grouting.

8.4.7 Vertical or nearly vertical drain pipes shall be provided in the dam body at downstream side of the upstream seepage proof zone. The lower ends of these drain pipes shall extend to the longitudinal drainage galleries, while the upper ends shall extend to the upper galleries, or the dam crest (or beneath the spillway surface) for facilitating inspection and repair. The drains may be formed by pulling pipes from concrete, drilling holes, or embedding precast porous concrete pipes. These drains may have an inner diameter between 15 cm and 25 cm, and are spaced generally 2-3 m.

Water seeping into the formed drains may be collected in the lower longitudinal galleries, and then conveyed to a sump pit through gutters or pipes, where the water will be pumped or discharged by gravity flow to the downstream area out of the dam. The gutters usually have a cross-section of 30 cm×30 cm and a longitudinal bottom slope of 3‰. During construction, the drain pipes must be protected from blockage by concrete or other debris.

8.5 Concrete Materials and Dam Zoning

8.5.1 The materials comprising the dam concrete are cement, aggregate, water, pozzolan (mineral admixtures) and chemical admixtures, etc., which shall conform to the current national standards and industry standards.

Dam concrete shall meet the requirements of design strength, and also other various requirements of durability and low heat performance with respect to, wherever applicable, impermeability, freeze resistance, abrasion resistance, and corrosion resistance to suit the working condition and local climatic environment.

8.5.2 Dam concrete shall be zoned on the basis of its position and working conditions (refer to Figure 8.5.2).

Ⅰ—exterior surface concrete above the maximum upstream and downstream water levels;

II—exterior surface concrete in the upstream and downstream zones of fluctuating water levels;

III—exterior surface concrete below the minimum upstream and downstream water levels;

IV—dam base concrete;

V—interior concrete of the dam body;

VI—abrasion-resistant concrete for overflow surfaces, water release outlets, training walls, and gate piers.

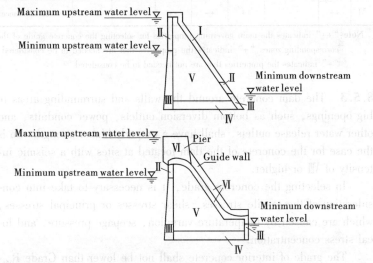

Figure 8.5.2 Zoning of dam concrete

The requirements on the properties of concrete at different dam zones are specified in Table 8.5.2.

Table 8.5.2 Properties of concrete at different dam zones

Zone No.	Strength	Impermeability	Freezing resistance	Abrasion resistance	Erosion resistance	Low heat of hydration	Maximum W/C ratio	Main properties in concrete zoning
I	+	-	++	-	-	+	+	Freezing resistance
II	+	+	++	-	+	+	+	Freezing resistance, cracking resistance

Continued to Table 8.5.2

Zone No.	Strength	Impermeability	Freezing resistance	Abrasion resistance	Erosion resistance	Low heat of hydration	Maximum W/C ratio	Mainproperties in concrete zoning
III	++	++	+	-	+	+	+	Impermeability, cracking resistance
IV	++	+	+	-	+	++	+	Cracking resistance
V	++	+	+	-	-	++	+	
VI	++	-	++	++	++	+	+	Abrasion resistance

Note: "++" indicates the main governing properties for selecting the concrete grade of the corresponding zones, "+" indicates the properties that are needed to be considered, "-" indicates the properties that are not needed to be considered.

8.5.3 The dam concrete around the walls and surrounding areas of big openings, such as bottom diversion outlets, power conduits, and other water release outlets, shall have a properly higher strength. So is the case for the concrete of the dam located at sites with a seismic intensity of VIII or higher.

In selecting the concrete grade, it is necessary to take into consideration of the tensile stresses, shear stresses or principal stresses, which are caused by temperature variation, seepage pressure, and local stress concentration.

The grade of interior concrete shall not be lower than Grade R_{90}-100, and the grade of exterior surface concrete at water passage portion shall not be lower than Grade $R_{28}250$.

8.5.4 The impermeability grade of concrete for seepage control shall be selected as specified in Table 8.5.4 on the basis of concrete location in the dam and the hydraulic gradient.

8.5.5 The grade of dam concrete for freezing resistance shall be determined on the basis of climate zone of a project, cycles of freezing and thawing, local ambient temperature on structure surface, moisture saturation degree of concrete, importance of the structural elements, easiness of repair and maintenance, etc. It shall also meet the requirements in *Design Specifications of Hydraulic Structures against Ice and Freezing Action* (SL 211).

Table 8.5.4   Minimum allowable values for impermeability grade of dam concrete

No.	Location	Hydraulic gradient	Impermeability grade
1	Interior part of dam		W2
2	Other locations	$i \leq 10$	W4
		$10 \leq i < 30$	W6
		$30 \leq i < 50$	W8
		$i \geq 50$	W10

Notes: 1. The symbol "i" indicates hydraulic gradient.
2. If the concrete is exposed to corrosive water, its impermeability grades shall be determined through special tests, and shall not be lower than W4.
3. The impermeability grade of concrete shall be determined according to the testing methods stipulated in *Design Specifications of Hydraulic Structures against Ice and Freezing Action* (SL 211). Depending on the initial time at which water pressure acting on the dam, it may also be determined by testing of concrete specimens at 90-day age.

8.5.6 The water-cement ratio of dam concrete should generally not exceed the values listed in Table 8.5.6 in order to meet the durability requirement.

Table 8.5.6   Maximum water-cement ratio

Climate zone	Dam concrete zone					
	I	II	III	IV	V	VI
Cold and severely cold zone	0.55	0.45	0.50	0.50	0.65	0.45
Temperate zone	0.60	0.50	0.55	0.55	0.65	0.45

8.5.7 In the corrosive water environment, corrosion-resistant cement shall be used in making the concrete. For the exterior concrete subjected to water level fluctuation or for underwater concrete, the water-cement ratio may be 0.05 less than the corresponding values given in Table 8.5.6.

8.5.8 In high-velocity flow area, low-fluidity and high-strength concrete with high resistance to abrasion and cavitation, or high performance silica fume concrete shall be adopted. If an abrasion-resistant lining is adopted, the lining shall be securely bonded with the concrete.

8.5.9 It is desirable to use no more than two different grades of concrete in one placement block, and the difference between the lowest and the highest grades should not exceed two grades. The minimum thickness of the concrete zones is 2-3 m.

9 Temperature Control and Crack Prevention

9.1 General

9.1.1 For high and medium dams, the design shall contain temperature control and crack prevention and shall also propose the temperature control criteria and crack prevention measures. For high dams, the finite element method is preferably to be used to obtain the temperature field and the thermal stresses in dams. For low dams, the temperature control and crack prevention design may follow the experiences of other projects with similar features.

9.1.2 Thermal cracks in a concrete gravity dam may be classified into three types: penetrated cracks, deep cracks, and surface cracks. Both the penetrated and deep cracks shall be avoided to occur in the dam. The surface cracks that occur on the upstream dam face or base concrete may possibly develop to become the penetrated or deep cracks, and hence they shall also be avoided.

9.1.3 Careful considerations shall be given to the arrangement of joints and concrete blocks and the control of temperature, and meanwhile measures shall be taken to improve or enhance the cracking resistance of the concrete. In the temperature control design, criteria for crack resistance of the concrete shall be proposed.

9.1.4 The data that shall be collected and analyzed for a dam site include the mean annual air temperature and its variation amplitude, the long-term mean monthly and mean 10-day air temperature, the amplitude, duration and frequency of sudden air temperature drops, the ground temperature of dam foundation, and the solar radiation, as well as the water temperature records of other reservoirs with similar features.

9.1.5 For high or medium dams, laboratory tests of concrete shall be performed to obtain the mechanical and thermal properties, ultimate tensile strain, creep, autogenous volumetric deformation, etc. For low dams, some of the essential tests may be performed as deemed necessary.

9.2 Temperature Control Criteria

9.2.1 "Foundation temperature difference" means the difference between the peak temperature and the final stable temperature of concrete in the foundation restraint zone with a height of $0.4L$ (L is the length of concrete placement block) above the foundation surface.

If the ultimate tensile strain of concrete at 28-day age in the foundation restraint zone is not less than 0.85×10^{-4}, the values of the allowable foundation temperature difference specified in Table 9.2.1 may be used for the concrete placement block uniformly placed in short time interval, with good and uniform quality, and with close deformation modulii of both the foundation rock and the concrete.

Table 9.2.1 **Allowable Foundation Temperature Difference** ΔT

(Unit: ℃)

Height above foundation surface, h	Length of concrete placement block, L				
	<17 m	17-20 m	20-30 m	30-40 m	40 m to full-length
$(0-0.2)L$	26-25	25-22	22-19	19-16	16-14
$(0.2-0.4)L$	28-27	27-25	25-22	22-19	19-17

9.2.2 The allowable foundation temperature difference of concrete shall be specially studied and justified for the following cases:

a The height-length ratio of the dam block being less than 0.5.

b Concrete placement blocks within the foundation restraint zone that are left exposed for a long time interval, or for water overflowing.

c Significant difference in the deformation modulus of foundation rock and the concrete.

d Concrete for pit backfilling and dental treatment in foundation, or concrete placement blocks on steep abutments.

e Test and observed data evidently indicating that the autogenous volumetric deformation of the concrete exhibits obvious and stable expansion or shrinkage.

f The temperatures of the dam blocks (such as the low level outlets, slotted dam blocks, and sluice floor slabs) in some case being lower than the final stable temperatures during construction and operation periods (the effect of such phenomenon shall be considered in design).

9.2.3 "Concrete temperature difference between upper concrete and lower concrete" means the difference between the maximum average temperature of the newly placed upper concrete and the average temperature of the lower old concrete, measured at the time of starting placing the upper concrete (the upper concrete and lower concrete are within one-fourth block length above and below the old concrete surface with concrete age longer than 28 days). When the upper concrete uniformly rises in short time interval placement for a total height of greater than 0.5 times the block length, the allowable temperature difference of the upper and lower concrete is generally between 15 ℃ and 20 ℃. When the upper concrete placement height is smaller than 0.5 times the block length, the value to be used for the allowable temperature difference of the upper and lower concrete shall be additionally studied.

9.2.4 Surface protection: When the daily average ambient temperature continuously drops totally to 6-9 ℃ or more within 2-3 days, protective measures shall be provided on the exposed surfaces of the concrete with an age less than 28 days (for temperate climate regions, concrete with an age up to 5 days is not prone to cracking). Stringent protection shall be applied to concrete surfaces at important locations, such as strong foundation restraint zone and the upstream dam face; ordinary protective measures shall also be provided at concrete surfaces in other locations.

The upstream dam face, the surfaces of the dam base concrete, and the surfaces of other important locations which may be exposed for longtime and be subject to the influences of annual variation of ambient air temperature, therefore, the protection duration and protective materials for such surfaces shall be determined by special study based on the local climate conditions.

The criteria for controlling peak temperatures of dam concrete in different seasons or in different months shall be proposed based on the local climate conditions.

9.2.5 During construction process, all the concrete blocks shall be raised as uniformly as possible, it is unsuitable that the height difference between the adjacent blocks exceeds 10-12 m and that the time interval for placing the adjacent blocks exceeds 30 days.

9.3 Temperature Control and Cracking Prevention Measures

9.3.1 According to the cracking resistance requirements for the concrete of high dams, the grade of base concrete at 28-day age should not be lower than R150-R200 (the corresponding ultimate tensile strain is 0.80×10^{-4}-0.85×10^{-4}). The grade of interior concrete at 90-day age shall not be lower than $R_{90}100$. For the water retaining concrete surface, the grade of concrete shall be determined based on the requirements of seepage control, cracking prevention, freezing resistance, construction conditions, etc.

For medium and low dams, the grade and tensile strain of concrete in the above-mentioned locations may be properly lowered according to the specific conditions at the site.

9.3.2 Suitable raw materials for producing the concrete shall be adopted to improve the concrete properties. The construction management and construction technology shall be improved to ensure the concrete being of high quality and the variation coefficient or the standard deviation of concrete strength under better control.

9.3.3 The transverse joints and longitudinal joints with respect to their locations shall be determined through technical and economic analysis and comparison based on the topographical and geological conditions of the dam foundation, the layout and cross-sectional dimension of the dam, the thermal stresses, the construction conditions, etc.

a The spacing of the transverse joints is normally 15-20 m. If the spacing of transverse joints is greater than 22 m or less than 12 m, justification is then required in the design.

b The spacing of the longitudinal joints is normally 15-30 m. If the length of a concrete block exceeds 30 m, temperature shall be strictly controlled. Concrete placing in full block length of high dams without longitudinal joints shall be adopted only after verification by specific study. Attention shall be paid to preventing the upstream face from deep cracking during the construction period or after reservoir impoundment.

9.3.4 The heat insulation measures for concrete surfaces shall attain the equivalent heat transfer coefficient (the calculation method is given in Annex F), to be determined in combination with the local climate conditions. The inlets and outlets of all galleries, water conveyance conduits, and shafts shall be sheltered or plugged in cold seasons and in the periods during sudden drop in air temperature.

9.3.5 Rational arrangement should be made on the concrete construction procedures and the concrete placement quantity in one whole year. Dam concrete within foundation restraint zone should be placed in low-temperature seasons. The height difference between the adjacent blocks and between the adjacent dam sections should be restricted. The concrete within the foundation restraint zone shall be placed to raise uniformly in short time intervals while placement of thin lift thickness in long time intervals shall be avoided. Preferably, concrete should be placed at nighttime in hot season; while concrete placement of dam body in winter should be avoided as far as possible in cold regions.

9.3.6 Suitable comprehensive measures shall be taken to reduce temperature rise due to heat of hydration of cement. They will include the following: to use low heat cement, to place low fluidity concrete, to add high-efficiency chemical admixture, to increase aggregate size, to optimize aggregate gradation, to add suitable pozzolan (mineral admixture), to control appropriate lift thickness and placement time interval between upper and lower lifts, and to provide water-cooling system. The lift thickness will be determined by calculation based on the temperature control criteria. Generally, the lift thickness should be 1.5-2.0 m for the concrete in the dam base area, and it may be greater for the concrete placed above the base concrete block so long as the temperature control criteria are met. In summer, the lift thickness should be reduced generally, but not less than 1.0 m thick. The normal time intervals shall be ensured and the placed concrete be cured with low-temperature natural river flow water.

9.3.7 The concrete placing temperature shall be reduced according to the design requirements. Reducing placing temperature may be accomplished by reducing the concrete temperature with measures adopted before the discharge of batched concrete from the plant; which are to spray water onto the coarse aggregate, to stockpile the aggregates in larger height, to take aggregate from the trough, to add ice and cold water for mixing concrete, to pre-cool the aggregates, to protect the pre-cooled aggregates from heat gaining, etc. Other measures are to strictly control the transport time of concrete and the exposure time of placed concrete before being covered, and to reduce the temperature gain during transport and placement.

9.3.8 Post-cooling of concrete by embedment of cooling pipe system in the dam body: in the initial period, the refrigerated water or cold river water may be used to reduce the peak temperature in concrete; in the intermediate period, the river water may be directly used to control the temperature difference between the interior and exterior concrete; in the final period, the refrigerated water or river water of low temperature may be used to reduce the temperature of the dam body down to the temperature suitable for joint grouting. The type of water to be utilized and the duration of cooling operation should be determined by means of analysis and computation.

During water cooling process, normally, the temperature difference between the dam concrete and the cooling water should not exceed 25 ℃, and the rate of temperature drop of concrete should not exceed 1 ℃/d.

9.3.9 For the reinforced concrete piers, walls and other portions using high-grade concrete, comprehensive measures, such as structural joints, temperature control, reinforcement arrangement and surface protection, should be taken to prevent and restrict cracking.

9.3.10 The temperature for contraction joint grouting is normally the final stable temperature of the dam body concrete. A slightly higher temperature may be selected for joint grouting in cold regions, if justifiable. Grouting of dam contraction joints shall be performed in accordance with *Construction Specifications on Cement Grouting for Hydraulic Structures* (SL 62—94). Preferably, grouting operation should be performed in cold season. For joint grouting for a solid dam section, if performed in hot season, surface insulation should be enhanced to avoid occurrence of overly high temperature near the dam surface.

9.3.11 For a slotted concrete gravity dam, surface insulation shall be maintained a long time during the construction period until closure of the wide slots.

9.3.12 For a hollow gravity dam, the temperature of the concrete below the hollow roof shall be reduced to the final stable temperature of the dam prior to the closure of the hollow roof. Efforts shall be made to reduce the peak temperature in newly placed concrete above the elevation of hollow roof.

10 Safety Monitoring Design

10.1 General

10.1.1 A concrete gravity dam shall be provided with necessary monitoring instruments or devices according to the dam class, height, and structural features, and according to the site geological conditions. The principal tasks of instrumentation and monitoring are:

a **To monitor the behaviors and safety of the various structures during periods of construction, initial reservoir impoundment and operation.**

b To check and verify the design, and to provide guidelines for construction and operation.

c To collect data for future use in design and research.

10.1.2 The scope of safety monitoring for a concrete gravity dam shall cover the dam body, dam foundation and abutments, plus the river banks or valley slopes in the vicinity of the dam, which may significantly affect the safety of the dam, and other appurtenant structures or equipment, which are directly associated with the dam safety.

10.1.3 The monitoring system shall be designed in accordance with the following principles:

a The monitoring system shall be able to reflect completely and accurately the behaviors of the dam and its foundation during periods of construction, initial reservoir impoundment and operation.

b Typical dam sections for special monitoring and for ordinary monitoring shall be selected in accordance with the dam height, geological conditions, structural features and the representativeness of similar dam sections. Measuring points shall be arranged at priority locations.

c The monitoring items and instrumentation layout shall be planned in an all-round way. For some important measuring points in the special monitoring sections or locations, two or more monitoring methods may be adopted; for important physical quantity to be measured at key locations, the instruments may have standby sets. The items of monitoring and layout of instruments shall be well selected in conjunction with the consideration of the major factors that will affect the safety of the project.

d The measuring instruments and equipment to be selected shall be reliable and stable in their performance, capable of long-term working in tough conditions. The measuring range and accuracy of the instruments shall be sufficient to meet the monitoring requirements. For the important features that need to be monitored permanently, the instruments shall be easy for replacement.

e Instrumentation using new technology should be adopted and they should provide means for possible updating in subsequent stages.

f Automated data acquisition and processing system should be adopted for measuring important points of gravity dams of Classe 1 and Class 2. Such system may be provided for gravity dams of Class 3, if considered necessary. Anyhow, manual measurements should also be available at locations using automated monitoring system.

10.1.4 Attention shall be paid to the following aspects in design of the monitoring system:

a The monitoring galleries and observation stations shall be reasonably arranged in conjunction with the dam structural design.

b Instrumentation system shall be provided with good support facilities, easy access, adequate lighting, reliable moisture-proof, wind protection, drainage, thermal insulation and security.

c The instruments shall be embedded and installed with minimal interference with construction operation. The instruments and cables shall be reliably protected.

d The predicted varying ranges of measured values for the major monitoring items should be given on the basis of results obtained from numerical computation and model tests. For the special monitoring dam section of Classe 1 and Class 2 gravity dams, alert values of the displacements and uplift pressures should be indicated preferably.

e **Special attention shall be paid to the design of safety monitoring during construction and during initial reservoir impoundment for timely acquiring the basic reference values of the**

various measuring points for key monitoring items. A detailed monitoring program shall be established prior to initial impounding of reservoir. If the permanent monitoring instruments and equipment are not completely installed or they are not ready to function before the initial reservoir impoundment, temporary monitoring facilities shall be correspondingly installed.

f According to the specific conditions of a project, technical specifications for the instrumentation and monitoring system should be formulated.

10.1.5 The design of instrumentation and monitoring system for concrete gravity dams shall conform to the provisions in *Specification on Instrumentation and Monitoring System of Concrete Dams* (SDJ 336).

10.2 Monitoring Items and Highlights for Instrument Layout

10.2.1 Safety monitoring for a concrete gravity dam consists of visual inspection and instrumentation.

10.2.2 Visual inspection shall meet the following requirements:

a The gravity dam and its appurtenant structures shall be periodically inspected from construction period to operation period. The frequency of the visual inspection shall be increased during periods of initial impounding of reservoir, rapid rise and drawdown of reservoir level, large flood releasing, felt earthquakes and other special events, if any.

b If any damages are found to the dam and its appurtenant structures, or any abnormal indicative phenomena are observed at the valley slopes near the dam, on groundwater level and on foundation seepage, they shall be immediately reported. Their causes shall be identified and remedial measures shall be studied for action.

c If possible, TV image monitoring system may be established for the important locations of Class 1 and Class 2 concrete gravity dams.

10.2.3 The monitoring items are categorized into two kinds, i.e. the routine monitoring items and special monitoring items, and all shall conform to the following provisions:

a The routine monitoring items may be determined according to Table 10.2.1 with respect to the dam class and with reference to the project features and geological conditions. Routine instrumentation shall be focused on monitoring the deformation and foundation seepage.

Table 10.2.1 **Routine monitoring items**

No.		Monitoring item	Dam class		
			1	2	3
1	Deformation	1) Dam displacement	●	●	●
		2) Tilt	●	○	
		3) Joint	●	●	○
		4) Crack	●	●	●
		5) Foundation displacements	●	●	●
2	Seepage	1) Seepage quantity and pumped water in gallery	●	●	●
		2) Uplift pressure and seepage pressure at deep part of foundation	●	●	●
		3) Seepage pressure in dam body	○	○	
		4) Seepage through dam abutments	●	●	●
		5) Water quality analysis	●	●	○
3	Stress, strain and temperature	1) Stress	●	○	
		2) Strain	●	○	
		3) Concrete temperature	●	●	○
		4) Foundation temperature	●	○	

Continued to Table 10.2.1

No.		Monitoring item	Dam class		
			1	2	3
4	Ambient conditions	1) Upstream and downstream water levels	●	●	●
		2) Air temperature	●	●	●
		3) Precipitation	●	●	●
		4) Reservoir water temperature	●	○	
		5) Sedimentation at upstream dam face	●	○	
		6) Downstream scouring	●	○	
		7) Ice and freezing	○		

Notes: ● means mandatory items; ○ means optional items.

b The special monitoring items shall be determined with respect to the dam class and its significance, the structural features, the construction technology and the geological conditions, and with reference to the following items for selection:

1) To monitor stability of river bank slopes near the dam.
2) To monitor stability of underground caverns.
3) To monitor the earthquake response.
4) To monitor the hydraulic characteristics.
5) Others.

10.2.4 The deformation monitoring instruments and their layout shall meet the following requirements:

a Tensioning wire alignment, vacuum laser alignment method and plumbline method should be used to measure the horizontal displacements of the dam and its foundation. When the dam axis is short and the atmospheric conditions are favorable, the collimation method or atmospheric laser alignment method may also be adopted to measure the horizontal displacement of the dam. An inverted plumbline shall be installed at each end of the alignment line as the base points.

b Plumblines are preferred for measuring the deflections of dam.

c Triangulation, collimation and intersection methods are suitable for measuring surface horizontal displacements of valley slopes and potential landslide areas close to the dam. Group of inverted plumblines, multiple-point extensometers, deflectometers or inclinometers may be used to measure horizontal displacements in deep zones.

d Groups of inverted plumblines, inclinometers or multiple-point extensometers are suitable instruments for monitoring the major tectonic fractures or weak structural planes within the dam foundation area.

e Precise leveling and vacuum laser alignment are suitable for measuring the vertical displacement of the dam and its foundation. Depending on the actual conditions, hydrostatic leveling may also be used. The starting reference points for precise leveling should be set on the bedrock of valley slopes near the dam. The benchmark for leveling shall be located downstream of the dam, free from the influence of the reservoir basin. The vacuum laser alignment system and the hydrostatic leveling instruments shall be located in the horizontal gallery in the dam, with reference points of vertical displacements set at both ends. The measuring points of vacuum laser alignment system and hydrostatic leveling should be arranged in conjunction with those for precise leveling.

f Precise leveling, settlement meters and multi-point extensometers are preferable instruments to measure the vertical displacements of valley slopes and potential landslide areas close to the dam. The trigonometric leveling method may also be used in mountainous areas. If necessary, the trigonometric leveling method may be made in conjunction with triangulation or trilateration surveys to form a three-dimensional network.

g Precise leveling, inclinometers, and hydrostatic leveling should be selected for measuring the inclinations of the dam and its foundation.

10.2.5 The layout of seepage monitoring instruments shall meet the following requirements:

a According to the size of the project, the foundation conditions and the designed seepage control measures, one to two longitudinal monitoring sections and at least three transverse monitoring sections should be provided to measure the uplift pressure in the foundation by piezometers, standpipes or seepage pressure meters.

1) The longitudinal monitoring section should be located at the first drainage line, usually with one measuring point provided in each dam monolith; however, the measuring points should be appropriately increased in dam monoliths seated on foundation with complicated geological conditions.

2) The transverse monitoring sections should be selected in the highest dam monolith, in the abutment dam monolith, and in the monoliths on terraced slope with complicated geological conditions. The transverse monitoring sections should be spaced 50-100 m and the spacing may be appropriately increased for dam foundation with simple geological condition. The measuring points should not be less than 3 at each monitoring section. If necessary, the measuring points shall be arranged on the upstream side of the grout curtain. When a downstream grout curtain is provided, the measuring points shall be located on its upstream side.

b In order to measure the seepage pressure in deep foundation zone, piezometers, standpipes or seepage pressure meters may be installed corresponding to the geological conditions and major geological defects. If a large fault or a highly pervious zone exists, the measuring points shall be installed along the possible seepage flow direction.

c In order to measure the seepage pressure in the dam body, one row of seepage pressure meters should be installed in the dam concrete at such locations: in between the upstream face and drain pipes, along the midway between two adjacent drain pipes in the dam, at the horizontal construction joints in the river flow direction, and also at the midway of upper and lower horizontal construction joints.

d In order to measure the seepage through the abutments, 2-3 monitoring sections may be provided downstream of the grout curtain along the seepage flow line in each abutment, and not less than 3 measuring points shall be installed for each monitoring section. The holes for piezometers shall be drilled deep enough to reach the highly pervious layer and well below the original groundwater table before the dam construction. If necessary, a few measuring points shall also be installed upstream of the grout curtain.

e In order to measure the seepage quantity through the dam body, foundation, riverbed and abutments, the measuring weirs should be installed at certain intervals in the foundation gallery drainage gutters. When excessive seepage is caused by the concrete defects, cold joints and cracks, it shall be collected and then measured with volumetric method. For the drain holes with excessive seepage, this volumetric method is better to be used for each individual hole.

f Water quality analysis shall be conducted on water samples taken periodically from representative drain holes or from monitoring holes in the abutments. The results of water quality analysis shall be compared with those of reservoir water. If the water sample is erosive or contain some exudation, further sampling for comprehensive analysis shall be performed.

10.2.6 The layout of instruments for measuring stresses, strains, and temperatures shall meet the following requirements:

a Along the centerline in a special monitoring dam monolith, one monitoring section perpendicular to the dam axis of the block and one or more horizontal monitoring sections shall be provided.

b The stress and strain measuring devices should be installed mainly on the special monitoring sections (vertical and horizontal). If necessary, certain number of measuring points shall also be selected at representative locations, such as large openings, periphery of galleries, and the vicinity of dam-foundation interface, or at other locations where stress conditions are complicated.

c Temperature monitoring instruments may be installed in the special monitoring dam block. The measuring points shall be arranged according to the dam temperature field, in combination with the dam surface temperature and foundation temperature. It is advisable that the instrumentation for the temporary temperature monitoring during construction may be used for permanent temperature monitoring during operation.

d In the special monitoring dam monoliths, jointmeters may be installed at different elevations in the grouted longitudinal and transverse joints. For the abutment monoliths, jointmeters should be installed at the interface between concrete and bedrock according to the actual conditions. Crack meters should be embedded in the concrete where potential cracking may occur.

e The prestressed anchor rods or tendons shall be monitored by sampling measurement of their variation of stress conditions.

f For important reinforced concrete structures, reinforcement stressmeters may be installed to monitor the stresses in steel bars.

10.2.7 The monitoring of ambient conditions shall not only satisfy the requirements in Table 10.2.1 but also comply with the provisions of relevant national standards.

Annex A Formulas for Hydraulic Design

A.1 Weir Crest Shape, Pressure on Crest Surface and Bucket Radius

A.1.1 The WES exponential curve is employed to define the downstream quadrant of the free overflow weir crest shape. It may be expressed by Equation (A.1.1):

$$x^n = k H_d^{n-1} y \quad (A.1.1)$$

where H_d—weir crest shape design head, m, to be calculated as 75%-95% of the maximum head (H_{max}) on weir crest based on permissible negative pressure on the crest;

x—horizontal coordinates positive to downstream with the highest point of the crest as origin;

y—vertical coordinates positive downward;

n—exponent, variable in relation to upstream face slope, refer to Table A.1.1;

k—when $P_1/H_d > 1.0$, k is taken from Table A.1.1; if $P_1/H_d \leqslant 1.0$, k is taken as 2.0-2.2; P_1 is the relative weir height at upstream side (see Figure A.1.2-3).

Table A.1.1 Parameter of crest curve

Upstream face slope $\Delta y/\Delta x$	k	n	R_1	a	R_2	b
3 : 0	2.000	1.850	$0.50H_d$	$0.175H_d$	$0.20H_d$	$0.282H_d$
3 : 1	1.936	1.836	$0.68H_d$	$0.139H_d$	$0.21H_d$	$0.237H_d$
3 : 2	1.939	1.810	$0.48H_d$	$0.115H_d$	$0.22H_d$	$0.214H_d$
3 : 3	1.873	1.776	$0.45H_d$	$0.119H_d$	—	—

A.1.2 Each of the following three curves may be used to define the upstream quadrant of the free overflow weir crest:

a Two-circular-arc compound curve, as shown in Figure A.1.2-1. Parameters such as R_1, R_2, k, n, a, and b are taken from Table A.1.1.

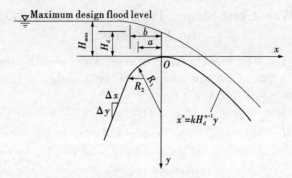

Figure A.1.2-1 Crest shape with two-circular-arc curve as upstream quadrant and exponential curve as downstream quadrant

b Three-circular-arc compound curve, with vertical upstream face, as shown in Figure A.1.2-2.

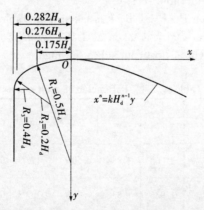

Figure A.1.2-2 Crest shape with three-circular-arc curve as upstream quadrant and exponential curve as downstream quadrant

c Elliptical curve, to be defined by Equation (A.1.2):

$$\frac{x^2}{(aH_d)^2} + \frac{(bH_d - y)^2}{(bH_d)^2} = 1.0 \qquad (A.1.2)$$

where aH_d, bH_d — major and minor semi-axis of the elliptic curve (when $P_1/H_d \geqslant 2$, $a = 0.28$-0.30, $a/b = 0.87 + 3a$; when $P_1/H_d < 2$, $a = 0.215$-0.28, $b = 0.127$-0.163; when P_1/H_d takes low values, a and b shall take the corresponding low values).

When corbel is provided on the upstream face of the weir as shown in Figure A.1.2-3, the Equation is applicable only under the condition of $d > H_{max}/2$.

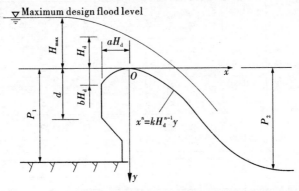

Figure A.1.2-3 Crest shape with elliptical curve as upstream quadrant and exponential curve as downstream quadrant, with corbel on upstream face

A.1.3 The design head H_d for WES weir is also related to permissible negative pressure on the wire surface. The minimum relative pressure h_{min}/H_d nearby the crest is dependent on the relative water head H_{max}/H_d, the relative weir height P_1/H_d and the relative downstream weir height P_2/H_d, as shown in Table A.1.3.

Table A.1.3 Minimum relative pressure h_{min}/H_d near WES weir crest

H_d/H_{max}	H_{max}/H_d	h_{min}/H_d														
		$\dfrac{P_1}{H_d} \geq 1.33$	$P_1/H_d=0.5$				$P_1/H_d=0.2$					$P_1/H_d=0.1$			
			P_2/H_d				P_2/H_d					P_2/H_d			
			0.5	1.0	1.5	3.0	0.2	0.4	0.6	1.2	2.2	0.1	0.2	0.8	1.1
0.60	1.67	−1.00	−0.02	−0.27	−0.48	−0.74	0.57	0.28	−0.18	−0.55	−0.72	0.85	0.34	−0.09	−0.48
0.65	1.54	−0.80	−0.01	−0.22	−0.42	−0.60	0.53	0.24	−0.16	−0.47	−0.56	0.79	0.24	−0.08	−0.42
0.70	1.43	−0.60	0	−0.15	−0.30	−0.41	0.48	0.21	−0.14	−0.38	−0.40	0.76	0.23	−0.07	−0.37
0.75	1.33	−0.45	0.02	−0.12	−0.23	−0.31	0.44	0.19	−0.10	−0.27	−0.27	0.71	0.20	−0.06	−0.30
0.775	1.29	−0.40	0.03	−0.09	−0.19	−0.26	0.43	0.18	−0.09	−0.24	−0.24	0.67	0.20	−0.05	−0.27
0.80	1.25	−0.30	0.05	−0.07	−0.16	−0.20	0.41	0.18	−0.07	−0.20	−0.20	0.65	0.20	−0.04	−0.24
0.825	1.21	−0.25	0.06	−0.04	−0.12	−0.16	0.39	0.18	−0.05	−0.16	−0.16	0.63	0.20	−0.03	−0.20
0.85	1.18	−0.20	0.07	−0.03	−0.11	−0.15	0.37	0.17	−0.04	−0.14	−0.14	0.62	0.20	−0.02	−0.18
0.875	1.14	−0.15	0.08	−0.02	−0.10	−0.12	0.36	0.17	−0.02	−0.11	−0.11	0.60	0.20	−0.01	−0.16
0.90	1.11	−0.10	0.08	0	−0.08	−0.09	0.35	0.17	0	−0.08	−0.08	0.57	0.20	0	−0.13
0.95	1.05	−0.05	0.10	0.02	−0.03	−0.04	0.33	0.16	0.03	−0.04	−0.04	0.55	0.20	0.03	−0.07
1.0	1.0	0	0.11	0.05	0	0	0.22	0.17	0.05	0	−0.01	0.52	0.20	0.03	−0.04

A.1.4 When breast wall is provided and the downstream quadrant of the ogee weir crest is a parabolic curve, and if the ratio of the maximum water head above the opening centerline to the opening height (H_{max}/D) is larger than 1.5, or if it is still belong to orifice flow even though the gate is fully opened, such curve may be defined by Equation (A.1.4):

$$y = \frac{x^2}{4\varphi^2 H_d} \quad (A.1.4)$$

where H_d—weir crest shape design head, m, normally taken as 75% to 95% of height from the centerline of the orifice to the maximum flood level;

φ —velocity coefficient in the contraction section, generally taken as 0.96; if an emergency or bulkhead gate slot is provided upstream the orifice, φ is taken as 0.95.

Other symbols are illustrated in Figure A.1.4.

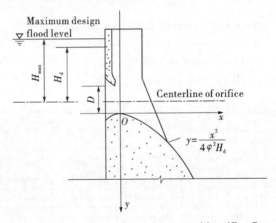

Figure A.1.4 Curve for weir crest with orifice flow

The upstream quadrant of the crest may be defined by either a single circular curve, or a two-circular-arc compound curve, or an elliptical curve.

A.1.5 The bucket radius shall be determined in conjunction with the downstream energy dissipation facilities. Different equations shall be used for different dissipation facilities.

a For flip bucket dissipator, the bucket radius may be calculated by Equation (A.1.5-1):

$$R = (4-10)h \qquad (A.1.5-1)$$

where R—bucket radius, i.e. the radius of reveres curve;

h—flow depth at bucket invert when the gate is fully opened at maximum flood level, m.

When the flow velocity, v, in bucket is less than 16 m/s, the lower limit in Equation (A.1.5-1) may be applied for bucket radius. The higher the flow velocity is, the larger the bucket radius should be adopted, until reaching the upper limit.

b For roller bucket dissipator, the bucket radius R is related to K, the ratio of specific discharge to energy. The relation curve of E/R and K is presented in Figure A.1.5. E/R is generally selected at a range of 2.1-8.4. The K is calculated by Equation (A.1.5-2).

$$K = \frac{q}{\sqrt{g}E^{1.5}} \qquad (A.1.5-2)$$

where K—the ratio of specific discharge to energy;

E—the gross energy head on the bucket invert, m;

q—specific discharge, m³/(s·m);

g—gravitational acceleration, m/s².

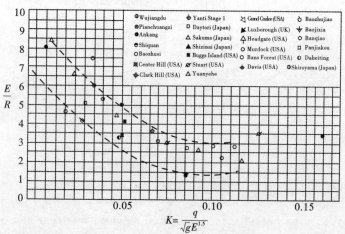

(Note: symbols and words in this figure indicating various example projects selected)

Figure A.1.5 Relation curve of E/R and K

c For the hydraulic jump (stilling basin) energy dissipators, the bucket radius may be calculated with reference to Equation (A.1.5-1).

A.2 Outline of Outlet Works through Dam Body

A.2.1 Outline of free flow openings through dam body

The typical profile of free flow opening, as shown in Figure A.2.1-1, consists of a shorter pressure flow portion and a longer free flow portion.

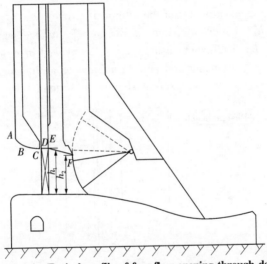

Figure A.2.1-1 Typical profile of free flow opening through dam body

The pressure flow portion comprises the entrance, bulkhead gate slot and constriction segments. Service gates are provided at the downstream constriction segments. The bulkhead gate will be a vertical-lift gate, whereas the service gate is mostly a radial gate.

The free flow portion consists of the straight line segment, the parabolic line segment and the reverse curve segment in the sequence from upstream to downstream.

a Entrance segment

The outline for each part of the entrance segment shall be designed as follows.

1) The roof curve of the entrance may be defined by compounded curve AB and BC, as described below:

Curve AB: elliptic curve is preferable. The major semi-axis may be the height of the opening at the end of the entrance segment and the minor semi-axis may be 1/3 of the major semi-axis. Therefore, the curve AB (as shown in Figure A.2.1-2) may be expressed by the following equation:

$$\frac{x^2}{(kh_1)^2} + \frac{y^2}{(kh_1/3)^2} = 1 \qquad (A.2.1-1)$$

where x, y—coordinates of the elliptical curves;

h_1—height of opening at the end of entrance segment;

k—coefficient, usually $k = 1$, however, in order to make the major and minor semi-axis as an integer, the value k may be taken as slightly larger than 1.0.

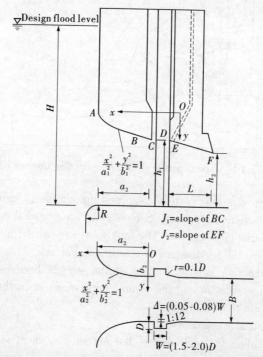

Figure A.2.1-2   Geometry of the pressure flow section

Line BC: A straight line tangent to curve AB at the point B, which may be preferably determined by Equation (A.2.1-2) as below:

$$\begin{cases} \dfrac{x}{3\sqrt{(kh_1)^2 - x^2}} = J_1 \\ \dfrac{x^2}{(kh_1)^2} + \dfrac{y^2}{(kh_1/3)^2} = 1 \end{cases} \quad (A.2.1\text{-}2)$$

where J_1—slope of tangent line BC, usually ranges from 1 : 4.5 to 1 : 6.5.

2) A single line of 1/4 ellipse defined by Equation (A.2.1-1) may also be used as roof curve to replace AB and BC.

3) Side curves: the side curves may be designed as a 1/4 ellipse. The equation of the ellipse curve may be expressed as:

$$\frac{x^2}{a_2^2} + \frac{y^2}{b_2^2} = 1 \quad (A.2.1\text{-}3)$$

where b_2—(0.22-0.27)B and $a_2 = 3b_2$, in which a_2 and b_2 are the major and minor semi-axis of the ellipse, respectively;

B —width of opening.

4) Invert shape: to be designed according to the actual conditions.

5) The height of upstream vertical face above the tangent point A shall not be less than the opening height at the downstream end of the entrance segment.

b Emergency or bulkhead gate slot segment

This segment consists of the sub-segments CD and DE. The gate slot with optimal shape and low incipient cavitation index shall be selected in the design. Between point C and point D, an open space is provided with a width of about 5 times the width of the sealing. Point C and point E shall be located at the same elevation.

c Roof constriction segment

EF is the roof constriction segment. It shall be shaped to ensure positive pressure at its entire length, and its roof slope should be slightly steeper than that of the segment BC, which may range from 1 : 6.0 to 1 : 4.0. For openings subject to high head, the greater value (1 : 4.0) is preferable, while for openings subject to low head or for water release structures of minor importance, the smaller value (1 : 6.0) may be applied. The ratio of the cross-sectional areas at the upstream end and downstream end of the roof constriction segment (A_2/A_1) may be determined according to the values used by the completed projects. When the seals of the emergency or bulkhead gate are provided downstream of the gate, attention shall be taken to provide air vents at the beginning of this segment.

d Free flow portion

The free flow portion starts from point F to downstream. The bottom vertical curve segment is normally designed to be a parabolic line as expressed by Equation (A.2.1-4):

$$y = \frac{g}{2(kv)^2 \cos^2\theta} x^2 + x\tan\theta \quad (A.2.1-4)$$

where θ—angle between horizontal and the line tangent to the parabola at the start point of the parabola (origin of coordinates x, y), if the preceding straight line segment is horizontal, $\theta = 0$;

v—average flow velocity of cross section at the start point, m/s;

g—acceleration of gravity, m/s^2;

k—factor of safety to avoid negative pressure, ranging from 1.2 to 1.6, generally $k = 1.6$.

The reverse curve segment should be designed as a single circular arc with a deflector or a lip at the end. The bucket lip shall be higher than the tail water level hereabout to ensure a free water jet; however, it is allowable to be slightly lower than the highest tail water level.

A.2.2 Outline of pressure flow opening through dam

The typical profile of a pressure flow opening through dam is as shown in Figure A.2.2. The shape of the entrance is basically the same as that of the free flow opening, except that the service gate is arranged at the exit of the opening. The emergency or bulkhead gate is still located downstream the entrance segment. The roof constriction segment is situated upstream of the service gate; a long horizontal pressure part is arranged to connect the gate slot segment at upstream and the exit constriction segment at downstream.

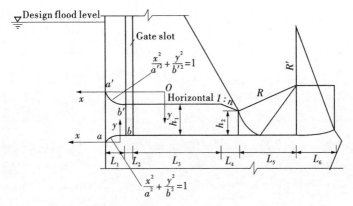

Figure A.2.2 Typical profile of pressure flow opening through dam

A.3 Calculation of Discharge Capacity and Aerated Water Depth

A.3.1 The discharging capacity of free overflow weir is calculated by Equation (A.3.1):

$$Q = Cm\varepsilon\sigma_s B\sqrt{2g}H_w^{3/2} \qquad (A.3.1)$$

where Q—discharge, m³/s;

B—net length of crest, m;

H_w—total head on crest, including velocity of approach head, m;

g—acceleration of gravity, m/s²;

m—discharge coefficient, refer to Table A.3.1-1;

C—correction coefficient related to upstream face slope, refer to Table A.3.1-2; when the upstream slope is vertical, C is taken as 1.0;

ε—pier contraction coefficient, determined by thickness of the pier and shape of the pier nose, $\varepsilon = 0.90$-0.95;

σ_s—submergence coefficient, to be determined based on degree of submergence of flow, for free flow, $\sigma_s = 1.0$.

Table A.3.1-1 Discharge coefficient

H_w/H_d	P_1/H_d				
	0.2	0.4	0.6	1.0	≥1.33
0.4	0.425	0.430	0.431	0.433	0.436
0.5	0.438	0.442	0.445	0.448	0.451
0.6	0.450	0.455	0.458	0.460	0.464
0.7	0.458	0.463	0.468	0.472	0.476
0.8	0.467	0.474	0.477	0.482	0.486
0.9	0.473	0.480	0.485	0.491	0.494
1.0	0.479	0.486	0.491	0.496	0.501
1.1	0.482	0.491	0.496	0.502	0.507
1.2	0.485	0.495	0.499	0.506	0.510
1.3	0.496	0.498	0.500	0.508	0.513

Notes: P_1 is weir height at upstream, m; H_d is weir crest shape design head, m, calculated as 75%-95% of the maximum head on crest (H_{max}).

Table A.3.1-2 Correction coefficient C related to upstream face slope

Slope $\Delta y/\Delta x$	P_1/H_d									
	0.3	0.4	0.5	0.6	0.7	0.8	0.9	1.0	1.2	1.3
3:1	1.009	1.007	1.005	1.004	1.003	1.002	1.001	1.000	0.998	0.997
3:2	1.015	1.011	1.008	1.006	1.004	1.002	1.001	0.999	0.996	0.993
3:3	1.021	1.015	1.010	1.007	1.005	1.002	1.000	0.998	0.993	0.988

A.3.2 Discharge capacity of orifice is calculated by Equation (A.3.2):

$$Q = \mu A_k \sqrt{2gH_w} \quad (A.3.2)$$

where Q—discharge, m³/s;

A_k—cross sectional area at exit, m²;

H_w—head on centerline of orifice in case of free flow, or difference between the upstream and downstream levels in case of submerged flow, m;

g—acceleration of gravity, m/s²;

μ—discharge coefficient of orifice or pipe flow; for orifice flow on weir crest with a breast wall, when $H_w/D = 2.0$-2.4 (D is opening height, m), $\mu = 0.74$-0.82; for short, low level pressure flow openings, $\mu = 0.83$-0.93; for long, low level pressure flow openings, μ must be determined after calculating the frictional and local form losses.

A.3.3 The air entraining water depth is calculated by Equation (A.3.3):

$$h_b = (1 + \frac{\zeta v}{100})h \qquad (A.3.3)$$

where h—flow depth, exclusive of aeration effect, m;

h_b—flow depth, inclusive of aeration effect, m;

v—average flow velocity in the calculated cross section, exclusive of aeration effect, m/s;

ζ—correction coefficient, generally taken as 1.0-1.4 s/m, to be determined based on flow velocity and the cross-section contraction condition, when the flow velocity is higher than 20 m/s, the greater value should be adopted.

A.4 Hydraulic Characteristics of Flip Bucket Energy Dissipation

A.4.1 Trajectory distance (refer to Figure A.4.1) may be estimated according to Equation (A.4.1):

$$L = \frac{1}{g}\left[v_1^2 \sin\theta\cos\theta + v_1\cos\theta\sqrt{v_1^2 \sin^2\theta + 2g(h_1 + h_2)} \right]$$

$$(A.4.1)$$

where L—jet trajectory distance, m, if the effect of flow concentration exists, the trajectory distance shall be multiplied by a reduction coefficient ranging from 0.90 to 0.95;

v_1—flow surface velocity at bucket lip, m/s, taken as 1.1 times the average flow velocity v at the bucket lip, i.e. $v_1 = 1.1v = 1.1\varphi\sqrt{2gH_0}$ (H_0 is the deference in elevations of reservoir water level and bucket lip, m);

θ—trajectory angle (°);

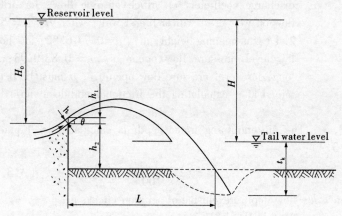

Figure A. 4. 1 Hydraulic characteristics of flip bucket energy dissipator

h_1 —vertical water depth on bucket lip, m, $h_1 = h\cos\theta$ (h is the average water depth on bucket lip, m);

h_2 —difference in elevations of bucket lip and riverbed, or of the bucket lip and the bottom of scour hole if the scour hole has been formed already, m;

φ —flow velocity coefficient of weir surface;

g —acceleration of gravity, m/s².

A. 4. 2 The maximum "water cushion" depth over the scour hole is estimated by Equation (A. 4. 2) as follows (refer to Figure A. 4. 1):

$$t_k = kq^{0.5}H^{0.25} \qquad (A.4.2)$$

where t_k—the water cushion depth, measured from tail water surface to the bottom of scour hole, m;

q—the specific discharge, m³/(s·m);

H—the head from reservoir level to tail water level, m;

k—the scouring coefficient, as listed in Table A. 4. 2.

Table A.4.2 Scouring coefficient k of bedrock

Scouring category		Hardly to scouring	Potentially scouring	Fairly scouring	Easily scouring
Joint and fissure	Spacing (cm)	>150	50-150	20-50	<20
	Development degree	Not developed, 1-2 sets of joints (fissures), regular	Moderately developed, 2-3 sets of joints (fissures), X-shaped, less regular	Developed, more than 3 sets of joints (fissures), X or asterisk shaped, irregular	Fairly developed, more than 3 sets of joints (fissures), disordered, cut and fragmented rocks
	Integrity	Blocky in Massive size	Blocky in Large size	Very Blocky	Extremely blocky
Formation characteristics of bedrock	Structure type	Integral	Masonry like structure	Interlocked structure	Disintegrated structure
	Feature of fissures	Mostly original or tectonic, mostly closed, extending short	Mostly tectonic, mostly closed, partially slightly open, occasional filling, sufficiently cemented	Mostly tectonic or weathered, mostly slightly open, partially open, some clay filling, insufficiently cemented	Mostly weathered or tectonic, slightly open or open, some clay filling, worse cemented
k	Range	0.6-0.9	0.9-1.2	1.2-1.6	1.6-2.0
	Average	0.8	1.1	1.4	1.8

Note: Applicable to cases with the jet impact angle ranging from 30° to 70°.

A.5 Hydraulic Characteristics of Energy Dissipation of Hydraulic-jump Type Stilling Basin

A.5.1 Formula for calculation of the basin length

a When Froude number of the incoming flow is $Fr \geqslant 4.5$ and when no appurtenance energy dissipation device is used, the length of apron for stilling basin is calculated by Equation (A.5.1-1) as below:

$$L = 6(h'' - h') \qquad (A.5.1-1)$$

where L—length of stilling basin, m;
  h'—conjugate water depth before jump, m;
  h''—conjugate water depth after jump, m.

b When $Fr > 4.5$, average velocity, v', of cross section at the beginning of stilling basin is higher than 16 m/s, and both the guide or chute blocks and end sill (without baffled blocks) are provided on the apron, the length of stilling basin is calculated by Equation (A.5.1-2) as below:

$$L = (3.2 - 4.3)h'' \qquad (A.5.1-2)$$

c When $Fr > 4.5$, average velocity, v', of cross section at the beginning of stilling basin is less than 16 m/s, and all the chute block, end sill and baffled piers are provided on the apron, the length of stilling basin is calculated by Equation (A.5.1-3) as below:

$$L = (2.3 - 2.8)h'' \qquad (A.5.1-3)$$

A.5.2 The pulsating pressure intensity on apron floor and end sill of the stilling basin may be estimated by Equation (A.5.2) as below:

$$P_m = \pm a_m \frac{v^2}{2g} \gamma_w \qquad (A.5.2)$$

where P_m—intensity of fluctuating pressure, acting on structure surface in the direction normal to the surface, kN/m^2;
  v—average flow velocity at computational section, m/s;
  a_m—coefficient of fluctuating pressure, ranging from 0.05 to 0.20, depending on flow velocity;
  γ_w—specific gravity of water, kN/m^3;
  g—acceleration of gravity, m/s^2.

A.5.3 The impact force, P_d, on baffled piers (or on chute blocks and end sill) may be estimated by Equation (A.5.3) as below:

$$P_\mathrm{d} = \pm k_\mathrm{d} \frac{v^2}{2g} \gamma_\mathrm{w} A_\mathrm{d} \qquad (A.5.3)$$

where P_d—impact force on baffled block (or on chute blocks and end sill), kN;

v—flow velocity in the vicinity of baffled piers, chute blocks and end sill, m/s, or approximately taken as the average flow velocity at the section of baffled piers, chute blocks and end sill;

A_d—projected area of upstream face of the pier, block or sill to flow direction, m^2;

k_d—resistance coefficient, ranging from 1.2 to 2.0 depending on location and shape of piers, blocks and sill, and the flow velocity as well;

γ_w—specific gravity of water, kN/m^3;

g—acceleration of gravity, m/s^2.

A.6 Prevention of Cavitation

A.6.1 The cavitation index of flow is calculated by Equation (A.6.1) as below:

$$\sigma_k = \frac{h_0 + h_\mathrm{d} - h_\mathrm{v}}{v_0^2/2g} \qquad (A.6.1)$$

where σ_k—cavitation index of flow, dimensionless;

h_0—water pressure at computational section including hydrodynamic pressure, in height of water column, m;

h_d—atmospheric pressure at the computational section, in height of water column, m, for different elevations, it is estimated as $(10.33 - \nabla/900)$, i.e. for each increment of 900 m in elevation from the sea level, the standard atmospheric pressure decreases by 1 m, ∇ is the elevation above sea level;

h_v—vapor pressure, in height of water column, m, referring Table A.6.1 for different water temperatures;

$v_0^2/2g$—average velocity head at computational section, m.

Table A.6.1　Vapor head of water versus water temperature

Water temperature(℃)	0	5	10	15	20	25	30	40
Vapor pressure h_v(m)	0.06	0.09	0.13	0.17	0.24	0.32	0.43	0.75

A.6.2 Unevenness means the abrupt offsets, depressions, or projections on the finished concrete surface subject to flowing water, which is not consistent with the design boundary shape. Such unevenness shall be strictly controlled during construction. The control criteria shall be determined in full consideration of the importance of boundary shape and the location concerned, the cavitation index of flow, the materials used for the surface and the possible continuous operation duration, etc. with reference to the criteria given in Table A.6.2.

Table A.6.2　Unevenness control criteria

Overflow drop (m)	Height of unevenness (mm)	Slope free from cavitation		
		Upstream slope	Downstream slope	Lateral slope
up to 20	up to 60	no limit	no limit	no limit
20-30	up to 30	no limit	no limit	no limit
	30-40	1 : 1	1 : 2	1 : 1
	40-60	1 : 1	1 : 2	1 : 1
30-40	up to 8	no limit	no limit	no limit
	8-10	no limit	1 : 2	1 : 1
	10-20	1 : 2	1 : 4	1 : 2
	20-40	1 : 6	1 : 10	1 : 3
	40-60	1 : 10	1 : 12	1 : 3

Continued to Table A.6.2

Overflow drop (m)	Height of unevenness (mm)	Slope free from cavitation		
		Upstream slope	Downstream slope	Lateral slope
40-50	up to 5	no limit	no limit	no limit
	5-10	1 : 4	1 : 8	1 : 2
	10-20	1 : 8	1 : 10	1 : 3
	20-40	1 : 12	1 : 14	1 : 3
	40-60	1 : 14	1 : 18	1 : 3
50-60	up to 3.5	no limit	no limit	no limit
	3.5-5	1 : 4	1 : 6	1 : 2
	5-10	1 : 10	1 : 14	1 : 3
	10-20	1 : 12	1 : 16	1 : 3
	20-40	1 : 16	1 : 18	1 : 3
	40-60	1 : 20	1 : 22	1 : 3
60-70	up to 2.5	no limit	no limit	no limit
	2.5-5	1 : 7	1 : 11	1 : 2
	5-10	1 : 14	1 : 18	1 : 3
	10-20	1 : 16	1 : 20	1 : 3
	20-40	1 : 20	1 : 24	1 : 3
	40-60	1 : 24	1 : 28	1 : 3
70-80	up to 10	1 : 20	1 : 24	1 : 3
	10-20	1 : 22	1 : 26	
	20-40	1 : 26	1 : 30	
	40-60	1 : 28	1 : 34	
80-90	10-20	1 : 28	1 : 32	1 : 4
	20-40	1 : 30	1 : 36	
	40-60	1 : 34	1 : 40	
90-100	10-20	1 : 32	1 : 38	1 : 4
	20-40	1 : 36	1 : 42	
	40-60	1 : 40	1 : 46	

A.6.3 Aeration device for cavitation prevention

a The aeration device for cavitation prevention shall meet the following requirements:

1) to ensure sufficient amount of air flows be introduced to the protected area, and to ensure the air concentration near the protection surfaces preferably not less than 3%-4%.

2) to have sufficient strength and reliability of its own.

3) to ensure smooth air supply by keeping the aeration cavity in stable condition and preventing plugging of aeration vents and slots, when releasing the design discharges and various low rates of flows.

b The aeration devices may be classified as:

1) Bottom aeration devices. Generally, the upturned ramps, vertical drops or slots will be provided on spillway surfaces to make the flow not contact to the bottom boundary so that cavity can be formed underneath the lower flow surface and air can be entrained through the cavity. This type of aerators causes less disturbance to the flow resulting in a smooth flow pattern, so as to protect the surfaces from cavitation damage.

2) Bottom and lateral aeration devices (also called abrupt expansion and abrupt drop device). The abrupt drop is provided at the bottom, meanwhile both lateral boundaries are abruptly expanded, which is favorable for arranging the upstream seals of gate subject to high pressure. Air will be introduced into the aeration cavity from three directions, so that the air concentration of flow is increased. However, because the pulsation action to the sidewalls due to lateral flow may increase, cavitation damage on sidewalls shall be carefully justified.

Annex B Formulas for Load Calculations

B.1 Hydrostatic Pressure at a Point on Dam Surface

$$P_w = \gamma_w H \quad (B.1)$$

where P_w—hydrostatic pressure at the computational point, kN/m^2;

γ_w—unit weight of water, kN/m^3, generally takes 9.81 kN/m^3, but for heavy silt-laden river, it shall be determined based on actual conditions;

H—head at the computational point, m, determined according to the height differential between the water level and the computational point.

B.2 Silt Pressure

The horizontal silt pressure acting on dam surface of unit length is calculated by Equation (B.2) as follows:

$$P_{sk} = \frac{1}{2}\gamma_{sb} h_s^2 \tan^2(45° - \frac{\varphi_s}{2}) \quad (B.2)$$

$$\gamma_{sb} = \gamma_{sd} - (1 - n)\gamma_w$$

where P_{sk}—silt pressure, kN/m;

γ_{sb}—buoyant unit weight of silt, kN/m^3;

γ_{sd}—dry unit weight of silt, kN/m^3;

n—porosity of silt;

h_s—silting depth in the vicinity of dam face, m;

φ_s—angle of internal friction of silt(°).

When the dam surface is inclined, the vertical silt pressure shall also be considered.

B.3 Uplift Pressure

B.3.1 Calculation of uplift pressure on dam-foundation interface

The distribution of uplift pressure along the base of a gravity dam shall be determined according to the following three conditions:

a When grout curtain and drains are provided in dam foundation, the uplift pressure at dam heel is the headwater H_1, at the dam toe is the tailwater H_2, and at the line of drains is $H_2 + \alpha (H_1 - H_2)$. The uplift pressure will vary linearly between the successive points along the base (refer to Figure B.3.1 (a), (b) and (c)).

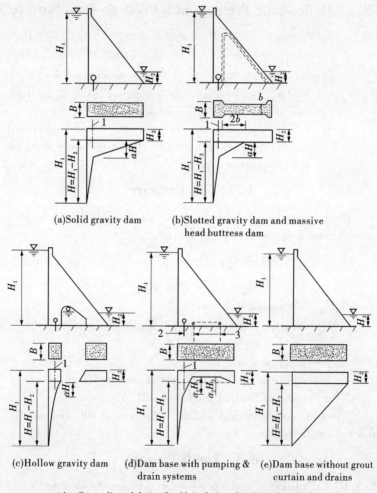

(a) Solid gravity dam (b) Slotted gravity dam and massive head buttress dam

(c) Hollow gravity dam (d) Dam base with pumping & drain systems (e) Dam base without grout curtain and drains

1—Center line of drain; 2—Main drains; 3—Secondary drains

Figure B.3.1 Distribution of uplift pressure along dam base

b When secondary downstream drains and pumping system are provided in dam foundation in addition to the grout curtain and the main drains near upstream face, the uplift pressure at the dam heel is the headwater H_1, at the dam toe is the tailwater H_2, and at the line of main drains and secondary drains is $\alpha_1 H_1$ and $\alpha_2 H_2$, respectively. Between two successive points, the uplift pressure will vary linearly as shown in Figure B.3.1(d).

c When no grout curtain and upstream drains are provided in dam foundation, the uplift pressure at the dam heel is the headwater H_1, and at the dam toe is the tailwater H_2. The uplift pressure will vary, as a straight line, from H_1 at the heel to H_2 at the toe as shown in Figure B.3.1(e).

d The values for all the coefficient of seepage pressure a, the uplift coefficient a_1 and the residual uplift coefficient a_2 may be taken as specified in Table B.3.1.

B.3.2 When drains are provided near the upstream face of the dam, the distribution of the uplift pressure on a computational plane within the dam may be determined according to Figure B.3.2. The coefficient of seepage pressure (a_3) at center line of drains in dam body may be taken as follows:

a For a solid gravity dam and the solid portion of a hollow gravity dam, $a_3 = 0.2$.

b For the continuous portion of a slotted gravity dam and massive-head buttress dam, $a_3 = 0.2$; whereas for the slot portion of the slotted gravity dam and the massive-head buttress dam, $a_3 = 0.15$.

B.3.3 When no drains are provided in dam body, the uplift pressure along the computational plane will vary linearly from full hydrostatic head H'_1 at the upstream face of the dam to zero or tailwater pressure H'_2 at the downstream face of the dam.

Table B.3.1 Seepage pressure and uplift coefficient along dam base

Dam type and location		Foundation treatment		
		(A) With grout curtain and drains	(B) With grout curtain, main and secondary drains, and pumping system	
Location	Type of dam	Seepage coefficient α	Uplift coefficient at main drains α_1	Residual uplift coefficient α_2
Sections on riverbed	Solid gravity dam	0.25	0.20	0.50
	Slotted gravity dam	0.20	0.15	0.50
	Massive-head buttress dam	0.20	0.15	0.50
	Hollow gravity dam	0.25	—	—
Sections on abutment	Solid gravity dam	0.35	—	—
	Slotted gravity dam	0.30	—	—
	Massive-head buttress dam	0.30	—	—
	Hollow gravity dam	0.35	—	—

Note: When dam foundation is provided with only drains but grout curtains, the seepage coefficient listed in column (A) shall be adequately increased to a certain extent.

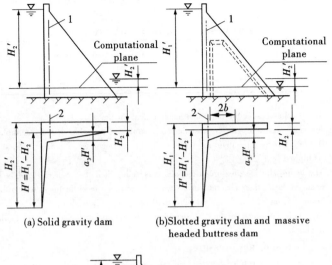
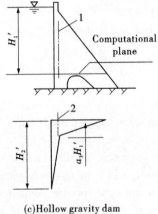

(c) Hollow gravity dam

1—Drain in dam body; 2—Centerline of drains

Figure B.3.2 Distribution of uplift pressure along a computational plane within dam

B.4 Ice Pressure

B.4.1 Static ice pressure

The static ice pressures on dam face of a unit length due to thermal expansion of the ice may be taken as given in Table B.4.1.

Table B.4.1 Hydrostatic ice pressure

Thickness of ice (m)	0.4	0.6	0.8	1.0	1.2
Standard value of static ice pressure (kN/m)	85	180	215	245	280

Notes:

1. The ice thickness is the maximum mean annual thickness.
2. For small reservoirs, the hydrostatic ice pressures in the table shall be multiplied by 0.87; while for large plain reservoirs with wide and open water surface, they shall be multiplied by 1.25.
3. The hydrostatic ice pressure listed above is valid when the reservoir level is barely changed during the icebound season; when the reservoir level fluctuates during the icebound season, the static ice pressure shall be specifically studied.
4. The static ice pressures may be determined by interpolation with respect to the ice thicknesses.

B.4.2 Dynamic ice pressure

The dynamic ice pressure acting on vertical dam face may be calculated by Equation (B.4.2) as below:

$$F_{bk} = 0.07 v d_i \sqrt{A f_{ic}} \qquad (B.4.2)$$

where F_{bk}—dynamic ice pressure induced by impact of ice to the structure, MN;

v—ice drift velocity, m/s, preferably to be determined based on actually measured data. When no measured data is available, the drift velocity of ice in river (canal) may be taken as the water flow velocity, and the drift velocity of ice in reservoir may be taken as 3% of the maximum wind speed encountered in this ice drifting period, but it should not be more than 0.6 m/s. For structures passing ice, the approaching velocity of drift ice in front of the structure may be adopted as the ice drift velocity;

A—area of drift ice, m^2, to be determined by the observed or collected data from the place in question or places nearby;

d_i—thickness of drift ice, m, to be taken as 0.7-0.8 times of the maximum thickness of ice at the places in question, and the larger value shall be adopted in the initial period of ice drifting;

f_{ic}—compressive strength of ice, MPa, preferably to be determined by testing, when no testing data is available, it may be taken as 0.3 MPa for reservoir, or as 0.45 MPa in the initial period of ice drifting and 0.3 MPa in the end period in river.

B.5 Pressure on the Bucket by Centrifugal Flow

B.5.1 The pressures on the reverse curve section of spillway or other water releasing structure by centrifugal force of flow are approximated to be distributed uniformly along the section in the value which may be calculated by Equation (B.5.1) as below:

$$P_{cr} = q\gamma_w v/R \qquad (B.5.1)$$

where P_{cr}—typical value of pressure by centrifugal force of flow, N/m^2;

q—unit width discharge along the section for the corresponding design case, m^3/(s·m);

v—average flow velocity of cross section at the lowest point of the reverse curve section, m/s;

R—radius of reverse curve, m;

γ_w—unit weight of water, kN/m^3.

B.5.2 The values of horizontal and vertical components of the resultant centrifugal force on the reverse curvature section of the spillway or other water releasing structures may be calculated by Equation (B.5.2-1) and Equation (B.5.2-2) as below:

$$P_{xr} = q\gamma_w v (\cos\varphi_2 - \cos\varphi_1) \qquad (B.5.2\text{-}1)$$
$$P_{yr} = q\gamma_w v (\sin\varphi_2 + \sin\varphi_1) \qquad (B.5.2\text{-}2)$$

where P_{xr}—value of horizontal component of the resultant centrifugal force per unit width, N/m;

P_{yr}—value of vertical component of the resultant centrifugal force per unit width, N/m;

φ_1, φ_2—angles as shown in Figure B.5.2, take the absolute values.

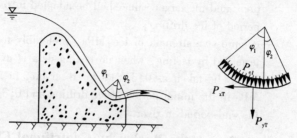

Figure B.5.2 Sketch of centrifugal force of flow on the reverse curve

B.6 Wave Pressure

B.6.1 The wave pressure on the vertical face of the structure shall be calculated in two cases according to the water depth behind the structure:

a When $H \geq H_{cr}$ and $H \geq \dfrac{L_m}{2}$, the distribution of wave pressure is shown in Figure B.6.1 (a). The wave pressure per unit width is calculated by Equation (B.6.1-1) through Equation (B.6.1-3) as belows:

$$P_{wk} = \frac{1}{4}\gamma_w L_m (h_{1\%} + h_z) \qquad (B.6.1-1)$$

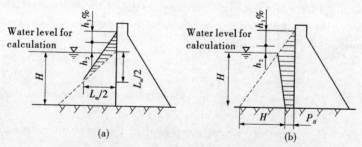

Figure B.6.1 Distribution of wave pressure on vertical face of water retaining structure

$$h_z = \frac{\pi h_{1\%}^2}{L_m} \operatorname{cth} \frac{2\pi H}{L_m} \qquad (\text{B.6.1-2})$$

$$H_{cr} = \frac{L_m}{4\pi} \ln \frac{L_m + 2\pi h_{1\%}}{L_m - 2\pi h_{1\%}} \qquad (\text{B.6.1-3})$$

where P_{wk}—wave pressure per unit width of the vertical face, kN/m;

γ_w—unit weight of water, kN/m^3;

L_m—average wave length, m;

$h_{1\%}$—height of wave with a accumulative frequency of 1%, m;

H—water depth behind the water retaining structure, m;

h_z—height from wave centerline to water level in question, m;

H_{cr}—critical water depth at which waves breach, m.

b When $H \geq H_{cr}$, but $H < \frac{L_m}{2}$, the distribution of wave pressure is shown in Figure B.6.1 (b). The wave pressure per unit width is calculated by Equation (B.6.1-4) and Equation (B.6.1-3) as belows:

$$P_{wk} = \frac{1}{2}[(h_{1\%} + h_z)(\gamma_w H + p_{lf}) + H p_{lf}] \qquad (\text{B.6.1-4})$$

$$p_{lf} = \gamma_w h_{1\%} \operatorname{sech} \frac{2\pi H}{L_m} \qquad (\text{B.6.1-5})$$

where p_{lf}—residual wave pressure at the bottom of the structure, kN/m^2.

B.6.2 Basic data necessary for calculating wave features such as height and length, etc.

a Maximum annual wind velocity means the maximum annual value of the 10-minute mean wind velocity at a height of 10 m above water surface. The wind velocity at a height of z (m) above water surface shall be calculated by multiplying the correction coefficient K_z as listed in Table B.6.2. The wind velocity not observed above the water surface but on land may be used only after being corrected with reference to the relevant data.

Table B.6.2 Wind velocity correction coefficient related to height

Height of z above water surface (m)	2	5	10	15	20
Correction coefficient K_z	1.25	1.10	1.00	0.96	0.90

b The length of wind field (effective fetch) is calculated as:

1) When the water surface covers wide areas along both sides of the wind direction, the fetch may be adopted as the straight distance from the computational point to the opposite shore.

2) When the water surface is locally contracted along wind direction and the width (b) of the surface at the contraction is less than 12 times of the computational wave length, the fetch may be adopted as $5b$, but it shall not be less than the distance from the computational point to the contracted location.

3) When the water surface along the wind direction is narrow or irregularly shaped, or small islands (or other obstacles) exist, a main fetch ray may be drawn from the computational point to meet the water boundary in the windward direction, then other rays at every 7.5° to both sides from the main fetch ray are drawn from the computational point to intersect the water boundary respectively. As shown in Figure B.6.2, D_0 is the water distance along the main fetch ray from the computational point to the opposite shore, D_i is the water distance along the ith ray from the computational point to the opposite shore, α_i is the angle between the ith ray and the main fetch ray, $\alpha_i = 7.5i$ (generally $i = \pm 1, \pm 2, \pm 3, \pm 4, \pm 5, \pm 6$), and meanwhile let $\alpha_0 = 0$, so the equivalent fetch length D may be calculated as:

$$D = \frac{\sum_i D_i \cos^2 \alpha_i}{\sum_i \cos \alpha_i} \qquad (i = 0, \pm 1, \pm 2, \pm 3, \pm 4, \pm 5, \pm 6)$$

(B.6.2)

c Mean water depth in wind field

The Mean water depth in wind field may generally be derived by plotting a topographical profile along the wind direction. The computational water level shall be in line with the hydrostatic level in the corresponding design cases.

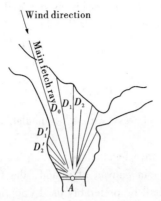

Figure B.6.2　Sketch for equivalent fetch length calculation

B.6.3 Calculation of the features of waves

　a　The features of wave should be calculated in the following three cases according to the conditions of the proposed reservoir:

　　1) For reservoirs in the plain or littoral area, the empirical equation derived by Putian Testing Station❶ should be used:

$$\frac{gh_m}{v_0^2} = 0.13 \text{th} \left[0.7 \left(\frac{gH_m}{v_0^2} \right)^{0.7} \right] \text{th} \left\{ \frac{0.001,8 \, (gD/v_0^2)^{0.45}}{0.13 \text{th} \left[0.7 (gH_m/v_0^2)^{0.7} \right]} \right\}$$

(B.6.3-1)

$$\frac{gT_m}{v_0} = 13.9 \left(\frac{gh_m}{v_0^2} \right)^{0.5}$$

(B.6.3-2)

where　h_m—mean wave height, m;

　　　　T_m—mean wave period, s;

　　　　v_0—calculated wind velocity, m/s;

　　　　D—fetch length, m;

　　　　H_m—mean water depth, m;

　　　　g—acceleration of gravity, 9.81 m/s².

　　2) For reservoirs in hilly or plain area, the Hedi Reservoir❷ em-

---

❶　In Guangdong Province.

❷　In Guangdong Province.

pirical equation should be preferably used (applicable to deep reservoirs, with $v_0 <$ 26.5 m/s and $D <$ 7.5 km):

$$\frac{gh_{2\%}}{v_0^2} = 0.006,25\, v_0^{1/6} \left(\frac{gD}{v_0^2}\right)^{1/3} \qquad (B.6.3\text{-}3)$$

$$\frac{gL_m}{v_0^2} = 0.038,6 \left(\frac{gD}{v_0^2}\right)^{1/2} \qquad (B.6.3\text{-}4)$$

where $h_{2\%}$—height of wave with a accumulative frequency of 2%, m;

L_m—mean wave length, m.

3) For reservoirs in canyon of inland, the Guanting Reservoir❶ empirical equation shall be preferably used (applicable to $v_0 <$ 20 m/s and $D <$ 20 km).

$$\frac{gh}{v_0^2} = 0.007,6 v_0^{-1/12} \left(\frac{gD}{v_0^2}\right)^{1/3} \qquad (B.6.3\text{-}5)$$

$$\frac{gL_m}{v_0^2} = 0.331 v_0^{-1/2.15} \left(\frac{gD}{v_0^2}\right)^{1/3.75} \qquad (B.6.3\text{-}6)$$

where h —$h_{5\%}$, the wave height at the accumulative frequency of 5% when $gD/v_0^2 =$ 20-250; or $h_{10\%}$, the wave height at the accumulative frequency of 10% when $gD/v_0^2 =$ 250-1,000,m.

b The relationship of the wave height h_P at the accumulative frequency P (%) to the mean wave height may be derived according to Table B.6.3.

c Relationship between mean wave length L_m and mean wave period T_m may be defined by Equation (B.6.3-7) as below:

$$L_m = \frac{gT_m^2}{2\pi} \text{th} \frac{2\pi H}{L_m} \qquad (B.6.3\text{-}7)$$

For deep-water waves, i.e. when $H \geqslant 0.5 L_m$, the above equation may be simplified to Equation (B.6.3-8) as below:

$$L_m = \frac{gT_m^2}{2\pi} \qquad (B.6.3\text{-}8)$$

---

❶ In Beijing Municipality.

Table B.6.3 Ratio of wave height at accumulative frequency P (%) to mean wave height

$\dfrac{h_m}{H_m}$	P (%)									
	0.1	1	2	3	4	5	10	13	20	50
0	2.97	2.42	2.23	2.11	2.02	1.95	1.71	1.61	1.43	0.94
0.1	2.70	2.26	2.09	2.00	1.92	1.87	1.65	1.56	1.41	0.96
0.2	2.46	2.09	1.96	1.88	1.81	1.76	1.59	1.51	1.37	0.98
0.3	2.23	1.93	1.82	1.76	1.70	1.66	1.52	1.45	1.34	1.00
0.4	2.01	1.78	1.68	1.64	1.60	1.56	1.44	1.39	1.30	1.01
0.5	1.80	1.63	1.56	1.52	1.49	1.46	1.37	1.33	1.25	1.01

B.6.4 The features of waves may be calculated based on B.6.3 above. But the calculated wind velocity should be determined in conformity with the following principles:

a When wave pressure is included in usual load combinations, the maximum annual wind velocity with a return period of 50 years is adopted as the calculated wind velocity.

b When wave pressure is included in unusual load combinations, the mean annual maximum wind speed is adopted as the calculated wind velocity.

Annex C Formulas for Stress Calculations of Solid Gravity Dams

C.1 Vertical Normal Stresses at Upstream and Downstream Faces

C.1.1 Vertical normal stress at upstream face (refer to Figure C.1.1):

$$\sigma_y^u = \frac{\Sigma W}{T} + \frac{6\Sigma M}{T^2} \quad (C.1.1)$$

where T —horizontal distance from upstream edge to downstream edge of computational plane, m;

ΣW—resultant vertical force above computational plane including uplift pressure (the same as below), positive in downward direction. For a solid gravity dam, a unit length of dam body along the axis is selected for calculation (the same as below);

ΣM—resultant moment of all the vertical and horizontal forces above computational plane about center of gravity of section, positive for moment produces compression at the upstream face.

C.1.2 Vertical normal stress at downstream face (refer to Figure C.1.1):

$$\sigma_y^d = \frac{\Sigma W}{T} - \frac{6\Sigma M}{T^2} \quad (C.1.2)$$

C.2 Shear Stresses at Upstream and Downstream Faces

C.2.1 Shear stress at upstream face:

$$\tau^u = (P - P_u^u - \sigma_y^u)m_1 \quad (C.2.1)$$

where m_1—slope of upstream face;

P—hydrostatic pressure at the upstream face of computational plane (sediment pressure also to be included, if any);

P_u^u—uplift pressure at upstream face of computational plane.

C.2.2 Shear stress at downstream face:

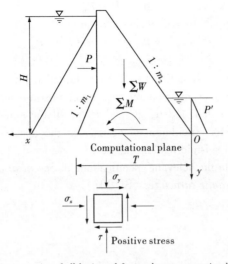

(ΣP is summation of all horizontal forces above computational plane, positive to the upstream direction)

Figure C.1.1　Sketch for stress calculation of solid gravity dams

$$\tau^d = (\sigma_y^d - P' + P_u^d)m_2 \quad (C.2.2)$$

where　m_2—slope of downstream face;

P'—hydrostatic pressure at downstream face of computational section (sediment pressure also to be included, if any);

P_u^d—uplift pressure at upstream face of computational section.

C.3　Horizontal Normal Stresses at Upstream and Downstream Faces

C.3.1　Horizontal normal stress at upstream face:
$$\sigma_x^u = (P - P_u^u) - (P - P_u^u - \sigma_y^u)m_1^2 \quad (C.3.1)$$

C.3.2　Horizontal normal stress at downstream face:
$$\sigma_x^d = (P' - P_u^d) + (\sigma_y^d - P' + P_u^d)m_2^2 \quad (C.3.2)$$

C.4　Principal Stresses at Upstream and Downstream Faces

C.4.1　Principal stresses at upstream face:
$$\sigma_1^u = (1 + m_1^2)\sigma_y^u - m_1^2(P - P_u^u) \quad (C.4.1\text{-}1)$$

$$\sigma_2^u = P - P_u^u \qquad (C.4.1\text{-}2)$$

C.4.2 Principal stresses at downstream face:
$$\sigma_1^d = (1 + m_2^2)\sigma_y^d - m_2^2(P' - P_u^d) \qquad (C.4.2\text{-}1)$$
$$\sigma_2^d = P' - P_u^d \qquad (C.4.2\text{-}2)$$

All the above Equation (C.2.1) to Equation (C.4.2-2) are applicable to the load combination with uplift pressure included. If the uplift effect is not included, P_u^u and P_u^d in all the equations shall be taken as zero for calculation of stresses at the upstream and downstream face. If seismic loading has to be considered, the calculations shall conform to the relevant specifications of *Specifications on Seismic Design of Hydraulic Structures* (SL 203).

Annex D Engineering Geological Classification and Mechanical Parameters of Rock Mass of Dam Foundation

Table D.1 Engineering geological classification of rock mass of dam foundation

Class	A. Hard rock mass (R_b >60 MPa)		B. Moderate-hard rock mass (R_b =60-30 MPa)		C. Soft rock mass (R_b <30 MPa)	
	Rock mass characteristics	Performance assessment of rock mass	Rock mass characteristics	Performance assessment of rock mass	Rock mass characteristics	Performance assessment of rock mass
I	A_1: rock mass of intact or blocky, extremely coarse lamellar or coarse lamellar structure, with an undeveloped-slightly developed structural plane of poor ductility and mostly closed, of an isotropic mechanical feature	Intact rock mass of high strength and better sliding and deformation resistance, special treatment for dam foundation not required, categorized as excellent foundation for high concrete dam				

Continued to Table D. 1

Class	A. Hard rock mass ($R_b > 60$ MPa)		B. Moderate-hard rock mass ($R_b = 60\text{-}30$ MPa)		C. Soft rock mass ($R_b < 30$ MPa)	
	Rock mass characteristics	Performance assessment of rock mass	Rock mass characteristics	Performance assessment of rock mass	Rock mass characteristics	Performance assessment of rock mass
II	A_{II}: good rock mass of blocky or sub-blocky, coarse lamellar structure, with a moderately developed structural plane, less weak structural planes or no wedges or prisms affecting stability of dam foundation or abutments	Rather intact rock mass with high strength and weak structural planes do not control stability of rock mass, with higher sliding and deformation resistance and not much special foundation treatment required, categorized as a good foundation for high concrete dam	B_{II}: structural characteristics of the rock mass are the same as that of A_I, and of an isotropic mechanical features	Intact rock mass of higher strength and higher sliding and deformation resistance, not much special foundation treatment required, categorized as a good foundation for high concrete dam		

Continued to Table D.1

Class	A. Hard rock mass (R_b >60 MPa)		B. Moderate-hard rock mass (R_b =60-30 MPa)		C. Soft rock mass (R_b <30 MPa)	
	Rock mass characteristics	Performance assessment of rock mass	Rock mass characteristics	Performance assessment of rock mass	Rock mass characteristics	Performance assessment of rock mass
III	A_{III1}: fair rock mass of a sub-blocky or medium to coarse lamellar structure, with the moderately developed structural plane, the rock mass distributed with weak structural planes of a low dip angle or high dip angle (dam abutments), or there are wedges or prisms affecting stability of dam foundation or dam abutments	Rather intact rock mass with localized poor integrity but higher strength, and the structural plane controls sliding and deformation resistance capability to a certain extent, special treatment required for those structural planes affecting the rock mass deformation and stability	B_{III1}: the rock mass structural characteristics are roughly the same as that of A_{II}	Rather intact rock mass with certain strength, and sliding and deformation resistance capability, dominated by the structural plane and the rock strength	C_{III}: rock strength of over 15MPa, rock mass of intact or extremely coarse lamellar structure, with a non-developed to moderately developed structural plane of the isotropic mechanic feature	Rather intact rock mass, and sliding and deformation resistance capability is dominated by the rock strength

Continued to Table D.1

Class	A. Hard rock mass (R_b >60 MPa)		B. Moderate-hard rock mass (R_b =60-30 MPa)		C. Soft rock mass (R_b <30 MPa)	
	Rock mass characteristics	Performance assessment of rock mass	Rock mass characteristics	Performance assessment of rock mass	Rock mass characteristics	Performance assessment of rock mass
III	A_{III2}: fair rock mass of an inter-bedded or inlaying fractured structure, with developed structural planes but scare structural plane of penetrability, structural planes characterized by a poor ductibility and mostly closed, with a better mosaic capability among rock mass	Poorer rock mass yet with higher strength, the sliding and deformation resistance is dominated by the structural plane and rock mass mosaic capability as well as shearing strength feature of the structural planes, special treatments are required for such structural planes	B_{III2}: fair rock mass of a sub-massive or medium to coarse lamellar structure, with moderately developed structural plane, mostly closed, better mosaic capability among rock mass, scarce penetrability structural plane	Rather intact rock mass yet poor in local places, sliding & deformation resistance are dominated by the structural plane and rock strength to certain extent		

Continued to Table D.1

Class	A. Hard rock mass (R_b >60 MPa)		B. Moderate-hard rock mass (R_b =60-30 MPa)		C. Soft rock mass (R_b <30 MPa)	
	Rock mass characteristics	Performance assessment of rock mass	Rock mass characteristics	Performance assessment of rock mass	Rock mass characteristics	Performance assessment of rock mass
IV	A_{IV1}: poor rock mass of an inter-bedded or lamellar structure, with rather developed or developed structural planes, obvious present in weak structural planes, wedges or prisms adversely affecting stability of the dam foundation and abutments	Poor rock mass, with sliding & deformation resistance obviously is dominated by structural plane and rock mass mosaic capability. Whether it can be used as foundation of a high concrete dam depends on the effectiveness of foundation treatments.	B_{IV1}: poor rock mass of an inter-bedded or lamellar structure, weak structural planes, wedges or prisms exist which is unfavorable to the stability of dam foundation (abutments)	Poor rock mass, with sliding & deformation resistance are obviously dominated by structural plane and rock mass mosaic capability, whether it can be used as a concrete high dam foundation shall be determined by foundation treatments.	C_{IV}: with rock strength of over 15 MPa, developed structural planes or a moderately developed structural planes with the rock strength of less than 15 MPa	Rather intact rock mass of low strength & poor sliding & deformation resistance, not suitable for use as a high concrete dam foundation. In case that such rock mass exists locally, special treatments will be required

Continued to Table D. 1

Class	A. Hard rock mass (R_b >60 MPa)		B. Moderate-hard rock mass (R_b =60-30 MPa)		C. Soft rock mass (R_b <30 MPa)	
	Rock mass characteristics	Performance assessment of rock mass	Rock mass characteristics	Performance assessment of rock mass	Rock mass characteristics	Performance assessment of rock mass
IV	A_{IV2}: poor rock mass of a fractured structure, with rather developed structural planes, mostly opened and sandwiched with debris and mud, and with a weak mosaic capability among the rock mass	Rather fractured rock mass with poor sliding & deformation resistance, not suitable for the high concrete dam foundation, in case of local presence of such rock mass, special treatments are required	B_{IV2}: poor rock mass of lamellar or fractured structure, with much developed structural planes of most openings and a poor mosaic capability among rock mass	Rather fractured rock mass with poor sliding & deformation resistance, not suitable for high concrete dam foundation, in case of local presence of such rock mass, special treatments are required		
V	A_V: very poor rock mass of a loose structure composed of the rock sandwiched with mud or rock encased by mud, characterized by the loose continuity medium	Fractured rock mass not suitable for the high concrete dam foundation. In case that such rock mass exists in certain locations of the dam foundation, special treatments will be required.	A_V: very poor rock mass of a loose structure composed of the rock sandwiched with mud or rock encased by mud, characterized by the loose continuity medium	Fractured rock mass not suitable for the high concrete dam foundation. In case that such rock mass exists in certain locations of the dam foundation, special treatments will be required.	A_V: very poor rock mass of a loose structure composed of the rock sandwiched with mud or rock encased by mud, characterized by the loose continuity medium	

Note: This classification applies to the concrete dam with a height over 70 m. R_b is the saturated uniaxial compressive strength.

Table D.2 Rock mass mechanical parameters of dam foundation

Rock class	Interface between concrete and dam foundation			Rock mass		Deformation modulus, E_0 (GPa)
	f'	C' (MPa)	f	f'	C' (MPa)	
I	1.50-1.30	1.50-1.30	0.85-0.75	1.60-1.40	2.50-2.00	40.0-20.0
II	1.30-1.10	1.30-1.10	0.75-0.65	1.40-1.20	2.00-1.50	20.0-10.0
III	1.10-0.90	1.10-0.70	0.65-0.55	1.20-0.80	1.50-0.70	10.0-5.0
IV	0.90-0.70	0.70-0.30	0.55-0.40	0.80-0.55	0.70-0.30	5.0-2.0
V	0.70-0.40	0.30-0.05	—	0.55-0.40	0.30-0.05	2.0-0.2

Notes: 1. f' and C' are shear-friction parameters whereas f is friction strength parameter.

2. Parameters here specified is applied to hard rock foundation, while for the soft rock, corresponding deduction will be applied based on the softening factor.

Table D.3 Mechanical parameters of structural planes, weak layers and faults

Classification	f'	C' (MPa)	f
Cemented structural plane	0.80-0.60	0.250-0.100	0.70-0.55
No filling structural plane	0.70-0.45	0.150-0.050	0.65-0.40
Cobles and debris	0.55-0.45	0.250-0.100	0.50-0.40
Rock debris with clay-filling	0.45-0.35	0.100-0.050	0.40-0.30
Clay-filled rock debris	0.35-0.25	0.050-0.020	0.30-0.23
Clay	0.25-0.18	0.005-0.002	0.23-0.18

Notes: 1. f' and C' are shear-friction parameters and f is friction strength parameter.

2. Parameters in the table is only limited to structural planes in hard rock.

3. For structural planes in soft rocks, corresponding deduction will be applied.

4. For the shear-friction of the cemented or no filling structural planes, either the maximum value or the minimum value will be selected based on the roughness of the structural plane.

Annex E Calculations of Stability against Sliding along Potential Failure Planes within Foundation

E.0.1 When sub-horizontal potential failure planes exist in deep zone of a dam foundation, stability analysis against sliding along deep-seated potential failure planes shall be performed. The potential failure planes may be grouped as single slip plane, double slip plane and multiple slip plane according to geological conditions.

Among these slip planes, the double slip plane is frequently encountered, as shown in Figure E.0.1.

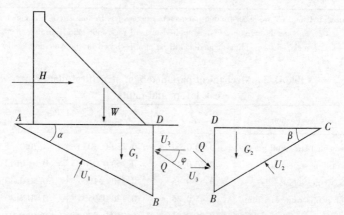

Figure E.0.1 Sketch of double slip plane

The factor of safety for stability against sliding along each of the slip planes shall be calculated either by shear-friction equation or by friction equation based on the "identical factor of safety method".

E.0.2 Calculation with shear-friction equation

Considering the stability of the wedges, the shear-friction stability safety factor of wedge ABD against sliding is expressed as:

$$K'_1 = \frac{f'_1 \left[(W + G_1)\cos\alpha - H\sin\alpha - Q\sin(\varphi - \alpha) - U_1 + U_3\sin\alpha \right] + c'_1 A_1}{(W + G_1)\sin\alpha + H\cos\alpha - U_3\cos\alpha - Q\cos(\varphi - \alpha)}$$

(E.0.2-1)

While shear-friction stability safety factor of the wedge BCD against sliding is expressed as:

$$K_2' = \frac{f_2'\left[G_2\cos\beta + Q\sin(\varphi+\beta) - U_2 + U_3\sin\beta\right] + c_2'A_2}{Q\cos(\varphi+\beta) - G_2\sin\beta + U_3\cos\beta}$$

(E.0.2-2)

where K_1', K_2'—shear-friction stability factors of safety against sliding;

W—vertical component of all loads (excluding uplift pressure, the same as below) acting on dam, kN;

H—horizontal component of all loads on dam, kN;

G_1, G_2—vertical forces by weight of rock wedges ABD and BCD, respectively, kN;

f_1', f_2'—coefficient of shear-friction on slip planes AB and BC, respectively;

c_1', c_2'—unit cohesion on slip planes AB and BC, respectively, kN;

A_1, A_2—areas of slip planes AB and BC, respectively, m^2;

α,β—angles between slip planes AB and BC and horizontal, respectively;

U_1, U_2, U_3—uplift pressures acting on slip planes AB, BC and BD, respectively, kN;

Q—interaction on plane BD;

φ—angle between the direction of interaction Q on BD and horizontal, which is selected after verification, conservatively $\varphi=0$.

Leting $K_1'=K_2'=K'$ and solving the equilibrium Equation (E.0.2-1) and Equation (E.0.2-2), the values Q and K' can be obtained.

E.0.3 Calculation with friction equation

For some dam blocks, if the shear-friction factor of safety K' derived by Equation (E.0.2-1) and Equation (E.0.2-2) cannot meet the requirements as specified in Clause 6.4.1 even after taking into account the proposed foundation treatment measures, the factor of safety against sliding may be calculated by friction Equation (E.0.3-1) and Equation (E.0.3-2). The allowable factor of safety for such case may be adopted only after careful justification.

Considering the stability of wedge ABD, we have:

$$K_1 = \frac{f_1 \left[(W + G_1)\cos\alpha - H\sin\alpha - Q\sin(\varphi - \alpha) - U_1 + U_3\sin\alpha \right]}{(W + G_1)\sin\alpha + H\cos\alpha - U_3\cos\alpha - Q\cos(\varphi - \alpha)}$$

(E.0.3-1)

Considering the stability of wedge BCD, we have:

$$K_2 = \frac{f_2 \left[G_2\cos\beta + Q\sin(\varphi + \beta) - U_2 + U_3\sin\beta \right]}{Q\cos(\varphi + \beta) - G_2\sin\beta + U_3\cos\beta}$$

(E.0.3-2)

where K_1, K_2—friction factors of safety against sliding;

f_1, f_2 —coefficients of friction on slip planes AB and BC respectivly.

Letting $K_1' = K_2' = K'$ and solving the system of Equation (E.0.3-1) and Equation (E.0.3-2), the values of Q and K will be achieved.

Since sliding along a single slip plane is simple, it is not necessary to list the required equations here. The sliding along multi-slip planes is more complicated to be given here. The value K for sliding along multi-slip planes may be derived by writing the equilibrium equation for each wedge in reference to the equations for the double slip plane and solve the general equations accordingly.

Annex F   Calculation of Temperature and Thermal Stresses of Dam during Construction

F.1   Calculation of Concrete Temperature

F.1.1   The temperature calculation of dam concrete at early stage is mainly aimed to obtain the peak temperatures of concrete after placement with respect to various possible temperature control measures, so as to judge whether the concrete temperatures satisfy the allowable values specified for foundation temperature difference, temperature difference between upper and lower concrete, temperature difference between interior and exterior of concrete, and interior peak temperature in dam, etc. In turn, to provide basic data for temperature control measures and temperature stress analysis. Generally, temperature calculation of concrete at early stage may be performed by finite difference method or practical calculation method (see F.1.4), whereas for those with complex boundary conditions, finite element method may also be used.

F.1.2   One-way finite difference method

$$T_{n,\tau+\Delta\tau} = T_{n,\tau} + \frac{a_c \Delta\tau}{\delta^2}(T_{n-1,\tau} + T_{n+1,\tau} - 2T_{n,\tau}) + \Delta\theta_\tau$$

(F.1.2-1)

where $T_{n,\tau+\Delta\tau}$ —the temperature of a computational point at the time step of calculation, ℃;

$T_{n,\tau}$ —the temperature of the computational point at the time step before calculation, ℃;

$T_{n-1,\tau}, T_{n+1,\tau}$ —temperatures of the upper and lower points adjacent to the computational point at the time step before calculation, ℃;

a_c —thermal diffusivity of concrete, m²/d;

δ —distance between two adjacent computational points, m;

$\Delta\tau$ —time interval of the time step, d;

$\Delta\theta_\tau$ —increment of adiabatic temperature rise of concrete in the calculation time step, ℃.

Stability criterion to be satisfied in calculation is $\dfrac{\alpha_c \Delta\tau}{\delta^2} \leq \dfrac{1}{2}$.

When the adiabatic temperature rise of concrete is expressed in Equation $\theta_\tau = \dfrac{\theta_0 \tau}{DN + \tau}$, the increment of adiabatic temperature may be expressed as:

$$\Delta\theta_\tau = \theta_0 \left(\dfrac{\tau + \Delta\tau}{DN + \tau + \Delta\tau} - \dfrac{\tau}{DN + \tau} \right) \quad \text{(F.1.2-2)}$$

$$\theta_0 = \dfrac{Q_0 W}{C_c \gamma_c}$$

Adiabatic temperature rise may also be expressed in Equation $\theta_\tau = \theta_0(1 - e^{-m\tau^b})$. In this case the increment is calculated by

$$\Delta\theta_\tau = \theta_0 \left[e^{-m\tau^b} - e^{-m(\tau + \Delta\tau)^b} \right] \quad \text{(F.1.2-3)}$$

where DN —the time required for generating half amount of heat of hydration in concrete, d;

m —the rate of hydration heat of cementitious materials for a given concrete, d^{-1};

b —a constant selected to make the mathematical curve fit the experimentally determined temperature rise curve of the concrete;

Q_0 —the final heat amount of hydration of cementitious materials, kJ/kg;

W —the content of cementitious material, kg/m^3;

C_c —the specific heat of concrete, kJ/(kg · ℃);

γ_c —the unit weight of concrete, kg/m^3.

F.1.3 Central finite difference method

$$T_{0,\tau+\Delta\tau} = T_{0,\tau} + \dfrac{2a_c \Delta\tau}{\delta^2} \left[\dfrac{1}{L_1 + L_2} \left(\dfrac{T_{1,\tau}}{L_1} + \dfrac{T_{2,\tau}}{L_2} \right) + \dfrac{1}{L_3 + L_4} \left(\dfrac{T_{3,\tau}}{L_3} + \dfrac{T_{4,\tau}}{L_4} \right) - T_{0,\tau} \left(\dfrac{1}{L_1 L_2} + \dfrac{1}{L_3 L_4} \right) \right] + \Delta\theta_\tau$$

(F.1.1-4)

where $T_{0,\tau+\Delta\tau}$ —the temperature at a computational point of a time step, ℃;

$T_{0,\tau}$ —the temperature at the computational point of the time step before calculation, ℃;

$T_{1,\tau}$, $T_{2,\tau}$ —temperatures from both the left and right points adjacent to the computational point of the time step before calculation, ℃;

$T_{3,\tau}$, $T_{4,\tau}$ —temperatures at the upper and lower points adjacent to the computational point of the time step before calculation, ℃;

δ —the mean distance between two adjacent computational points, m;

L_1, L_2 —ratios of the distance of the computational point from the left and right adjacent point to the mean distance δ, respectively;

L_3, L_4 —ratios of distance of the computational point from the upper and lower adjacent point to the mean distance δ, respectively.

When the central finite difference method is adopted in calculation, the stability criterion $\dfrac{\alpha_c \Delta \tau}{\delta^2} \leq \dfrac{1}{4}$ shall be satisfied. The concrete surface temperature may be treated as the Class Ⅲ boundary condition in general. When the concrete is cured with flowing water, the surface temperature may be taken as the average of the water temperature and the air temperature. For initial water cooling of the concrete by means of post-cooling, temperature calculation by finite difference method may be carried out in combination with the calculation for the initial water cooling, as described in details in Section F.2.

F.1.4 Practical calculation method

Whereas both the thermal conduction differential equation and the boundary conditions are linear, superposition techniques may be used to divide the complicated dissipation process of heat for a concrete block into four units for solution, as shown in Figure F.1.4-1.

a Equation for the mean temperature of the concrete block at early stage without the initial water cooling:

$$T_m = (T_u - T_s)E_1 + (T_p - T_s)E_2 + T_r + T_s \quad (F.1.4\text{-}1)$$

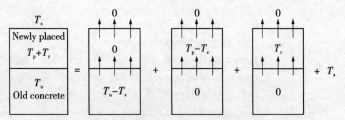

Figure F.1.4-1 Sketch showing the decomposition of temperature of a concrete block

In the case of uniform placement of a concrete lift with a short time interval, let $T_u \approx T_m$, the equation may then be simplified:

$$T_m = \frac{(T_p - T_s)E_2}{1 - E_1} + \frac{T_r}{1 - E_1} + T_s \quad (F.1.4\text{-}2)$$

$$E_1 = \frac{\sqrt{F_0}}{\sqrt{\pi}} \left(1 + e^{-\frac{1}{F_0}} - 2e^{-\frac{1}{4F_0}}\right) + P\left(\frac{1}{\sqrt{F_0}}\right) - P\left(\frac{1}{2\sqrt{F_0}}\right) \quad (F.1.4\text{-}3)$$

$$E_2 = \frac{\sqrt{F_0}}{\sqrt{\pi}} \left(4e^{-\frac{1}{4F_0}} - e^{-\frac{1}{F_0}} - 3\right) - P\left(\frac{1}{\sqrt{F_0}}\right) + 2P\left(\frac{1}{2\sqrt{F_0}}\right)$$

$$(F.1.4\text{-}4)$$

$$F_0 = \frac{a_c \tau}{l^2}$$

$$T_s = T_a + \Delta T$$

where T_m—the mean temperature of the covering concrete block, ℃;

T_p—placing temperature of covering concrete, ℃;

T_r—temperature rise of covering concrete due to heat of hydration, if it is calculated by time difference method, see Table F.1.4, ℃;

T_u—temperature of older concrete, ℃;

E_1—ratio of residual heat in the covering concrete after the action of heat absorbing from the older concrete and concurrent heat dissipating of the absorbed to its surface, as determined by Equation (F.1.4-3), or from Figure F.1.4-2;

a_c—thermal diffusivity of concrete, m²/d;

τ—time, d;

Table F.1.4 T_r calculated by time difference method

Time step	Time τ	Adiabatic temperature rise θ_τ	Increment of adiabatic temperature rise $\Delta\theta_i$	F_0	E_2	Temperature calculation			
						Time step 1	Time step 2	Time step 3	...
1	$\Delta\tau$	$\theta_{\frac{1}{2}\Delta\tau}$	$\Delta\theta_1 = \theta_{\frac{1}{2}\Delta\tau}$	$\dfrac{a_c \Delta\tau}{l^2}$	E_{21}	$E_{21}\Delta\theta_1$	$E_{22}\Delta\theta_1$	$E_{23}\Delta\theta_1$	...
2	$2\Delta\tau$	$\theta_{1\frac{1}{2}\Delta\tau}$	$\Delta\theta_2 = \theta_{1\frac{1}{2}\Delta\tau} - \theta_{\frac{1}{2}\Delta\tau}$	$\dfrac{2a_c \Delta\tau}{l^2}$	E_{22}		$E_{21}\Delta\theta_2$	$E_{22}\Delta\theta_2$	...
3	$3\Delta\tau$	$\theta_{2\frac{1}{2}\Delta\tau}$	$\Delta\theta_3 = \theta_{2\frac{1}{2}\Delta\tau} - \theta_{1\frac{1}{2}\Delta\tau}$	$\dfrac{3a_c \Delta\tau}{l^2}$	E_{23}			$E_{21}\Delta\theta_3$	...
...	...	...	...		...				...
			T_r			Σ	Σ	Σ	...

Note: E_2 value in this table may be determined either by Equation (F.1.3-4) or from Figure F.1.3-3 according to the value of F_0.

l —the lift thickness, m;

E_2 —ratio of residual heat in the covering concrete after action of dissipating of the heat from newly placed concrete both to the air and to the older concrete, as determined by Equation (F.1.4-4), or from Figure F.1.4-3;

T_a —ambient air temperature, ℃;

T_w —water temperature at inlet of cooling pipe, ℃;

T_s —surface temperature of the newly placed concrete, ℃;

ΔT —the temperature difference between the surface temperature of concrete and the ambient temperature.

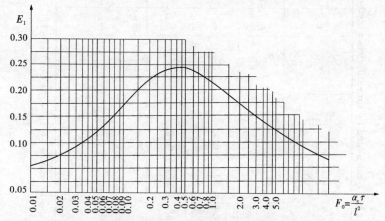

Figure F.1.4-2 Curve of ratio (E_1)

If the ambient temperature is constant, semi-infinite mass equation with heat source may be taken as an approximation of ΔT, i.e. the relation of ΔT versus τ may be obtained using Equation (F.1.4-5) and Equation (F.1.4-6), and then the required ΔT may be determined by interpolation at time τ.

$$\Delta T = \frac{\theta_\tau}{2 + \frac{\beta}{\lambda_c} x} \quad (\text{F.1.4-5})$$

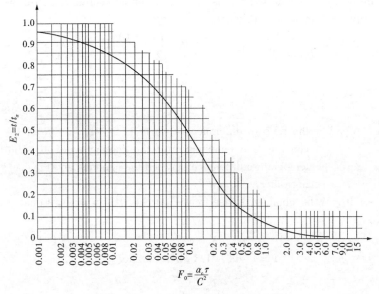

Figure F.1.4-3 Curve of ratio (E_2)

$$\tau = \frac{1}{6a_c}\left[\frac{x_2}{2} + \frac{2\lambda_c}{\beta}x - \left(\frac{2\lambda_c}{\beta}\right)^2 \ln\left(\frac{\beta}{2\lambda_c}x + 1\right)\right] \quad (F.1.4\text{-}6)$$

where x —the influence depth from the surface of the semi-infinite mass caused by surface heat loss at time τ, m;

θ_τ —temperature rise of heat of hydration at time τ, it is calculated by the equation derived by fitting the measured data of adiabatic temperature rise of concrete, ℃.

λ_c—the thermal conductivity of concrete, kJ/(m·h·℃).

It may also be determined according to measured data, i.e., $\Delta T = 2\text{-}5$ ℃ approximately (the low value should be taken for lower strength class of concrete). When one layer of straw sacks or other equivalent heat insulating material is covered on the top of concrete, $\Delta T \approx 10$ ℃ and when the concrete surface is cured by flowing water, $T_s = \frac{1}{2}(T_a + T_w)$.

b The following equation is used to calculate the mean temperature of a lift, when the heat loss during the initial cooling of post-cooling process shall be considered.

$$T_m = \frac{(T_p - T_s)E_2 X}{1 - E_1 X} + \frac{(T_w - T_s)E_2(1 - X)}{1 - E_1 X} + \frac{T_r}{1 - E_1 X} + T_s$$
(F.1.4-7)

$$X = f\left(\frac{a_c \tau}{D^2}, \frac{\lambda_c L}{C_w \gamma_w q_w}\right) \quad \text{(F.1.4-8)}$$

where X —the ratio of difference between mean temperature of the concrete and temperature of the cooling water to initial temperature difference between the concrete and the cooling water.

It may be obtained either from Figure F.1.4-4 or by Equation (F.1.4-9).

$$X = e^{-kF_0^s} \quad \text{(F.1.4-9)}$$
$$k = 2.08 - 1.174\xi + 0.256\xi^2$$
$$F_0 = \frac{a_c \tau}{D^2}$$
$$s = 0.971 + 0.148,5\xi - 0.044,5\xi^2$$
$$\xi = \frac{\lambda_c L}{C_w \rho_w q_w}$$
$$D = 2b = 1.21\sqrt{S_1 S_2}$$

where C_w —specific heat of water, kJ/(kg · ℃);
γ_w—unit weight of water, kN/m³;
q_w—rate of water flowing through a single cooling pipe, m³/h;
L—length of a single pipe, m;
D—the equivalent diameter of cooled cylinder, m;
S_1, S_2—the horizontal and vertical distance of cooling pipe, respectively, m.

When $F_0 = \dfrac{a_c \tau}{D^2} \leqslant 0.75$, the following equation may also be used:

$$X = e^{-k_1 F_0} \quad \text{(F.1.4-10)}$$
$$k_1 = 2.09 - 1.35\xi + 0.320\xi^2$$

Figure F.1.4-4 is given under the condition of $\dfrac{b}{c} = 100$. If $\dfrac{b}{c} \neq$

100, the thermal diffusivity of concrete a_c may be replaced by an equivalent coefficient of thermal diffusivity a'_c.

$$a'_c = \left(\frac{\alpha_1 b}{0.717,6}\right)^2 a_c \qquad (F.1.4\text{-}11)$$

$$\alpha_1 b = 0.926 e^{-0.031,4(\frac{b}{c}-20)^{0.48}}, \quad 20 \leq \frac{b}{c} \leq 130$$

where b —the radius of the cooled cylinder, m;
  c —the radius of the cooling pipe, m.

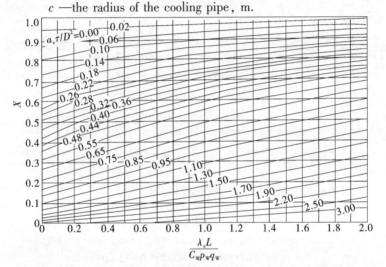

Figure F.1.4-4 Values of X in computation of pipe cooling of concrete

The equivalent coefficient of thermal conductivity may also be directly approximated by Equation (F.1.4-12).

$$a'_c = \frac{a_c \ln 100}{\ln \frac{b}{c}} \qquad (F.1.4\text{-}12)$$

F.1.5 Calculation of temperature of dam concrete in the late stage

In the late stage, heat of hydration of concrete will not be generated, and the dam is placed to a certain height. The heat exchange of concrete with the ambient environs takes place mainly through the upstream and downstream face as well as the top surface, which belongs to the temperature field without internal heat source. The temperature calculation is required to identify the temperature difference between the interior and exterior of concrete and in turn the thermal stresses induced, to estimate the rate of temperature drop and to determine the time required for joint grouting between dam blocks. Difference method or finite element method may be used for such a calculation.

F.2 Calculation of Temperature Reducing of Concrete by Pipe Cooling

F.2.1 Calculation of temperature reducing by pipe cooling in initial or early stage

a In initial stage, temperature of concrete by pipe cooling (with heat source) may be calculated by the following equation:

$$T_m = T_w + X(T_0 - T_w) + X_1 \theta_0 \qquad (F.2.1\text{-}1)$$

$$X_1 = f\left(\frac{a_c \tau}{b^2}, b\sqrt{\frac{m}{a_c}}, \frac{b}{c}, \frac{\lambda_c L}{C_w \gamma_w q_w}\right)$$

where T_m —mean temperature of concrete, ℃;

T_0 —initial temperature of concrete at the beginning of pipe cooling, ℃;

T_w —temperature of cooling water, ℃;

θ_0 —adiabatic temperature rise of concrete, ℃;

X —ratio of difference between mean temperature of the concrete and temperature of the cooling water to initial temperature difference between the concrete and the cooling water, determined either from Figure F.1.4-4 or by Equation (F.1.4-9).

X_1 —ratio of residual heat using cooling pipe system to dissipate heat in concrete, as shown in Figure F.2.1;

a_c —thermal diffusivity of concrete, m²/h;

τ —age of concrete, d;

b, D —radius and diameter of the cooled cylinder, m;

λ_c —thermal conductivity of concrete, kJ/(m · h · ℃);

L —length of a single cooling pipe, m;

C_w —specific heat of water, kJ/(kg · ℃);

γ_w —unit weight of water, kg/m³;

q_w —rate of water flowing through a single cooling pipe, L/min;

m —a constant representing the rate of hydration heat for a given concrete, d^{-1};

c —radius of cooling pipe, m.

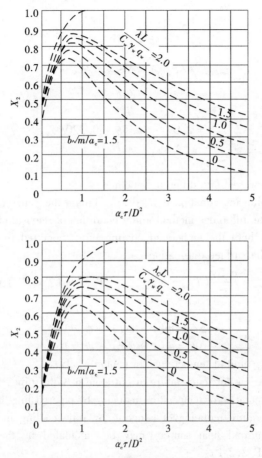

Figure F.2.1 Values of X_1 in computation of pipe cooling of concrete

b Equivalent heat conduction equation

Suppose that the adiabatic temperature rise of concrete is $\theta(\tau) = \theta_0(1 - e^{-m\tau})$, the heat loss shared by concrete surfaces and the cooling pipe may be calculated by solving the equivalent heat conduction Equation (F.2.1-2), in which the pipe cooling effect is considered, using finite element mesh.

$$\frac{\partial T}{\partial \tau} = a_c \left(\frac{\partial^2 T}{\partial x^2} + \frac{\partial^2 T}{\partial y^2} + \frac{\partial^2 T}{\partial z^2} \right) + (T_0 - T_w) \frac{\partial \phi}{\partial \tau} + \theta_0 \frac{\partial \psi}{\partial \tau}$$
(F.2.1-2)

$$\phi = e^{-p\tau}$$

$$\psi(\tau) = \frac{m}{m-p}(e^{-pt} - e^{-m\tau})$$

$$p = \frac{a_c k}{D^2}$$

$$k = 2.09 - 1.35\xi + 0.320\xi^2$$

$$\xi = \frac{\lambda_c L}{C_w \rho_w q_w}$$

c Forward finite difference method

The following equation is used to calculate the concrete temperature by finite difference method under condition of removal of heat from concrete by both the embedded cooling pipes operated in the initial stage and the surface dissipation.

$$T_{n,\tau+\Delta\tau} = T_{n,\tau} + \frac{a_c \Delta\tau}{\delta^2}(T_{n-1,\tau} + T_{n+1,\tau} - 2T_{n,\tau}) + (T_0 - T_w)\Delta\phi + \theta_0 \Delta\psi$$
(F.2.1-3)

$$\Delta\phi = \phi(\tau + \Delta\tau) - \phi(\tau)$$
$$\Delta\psi = \psi(\tau + \Delta\tau) - \psi(\tau)$$

where T_0 —initial temperature of concrete, ℃;

T_w —water temperature at entrance of water pipe, ℃.

F.2.2 Calculation of temperature reducing by pipe cooling in late stage

Temperature reducing of concrete by pipe cooling in late stage (without internal heat source) may be calculated by the following Equation (F.2.2-1):

$$T_m = T_w + X(T_0 - T_w) \quad \text{(F.2.2-1)}$$

where T_m —mean temperature of concrete, ℃;

X —ratio of difference between mean temperature of the concrete and temperature of the cooling water to initial temperature difference between the concrete and the cooling water, determined either from Figure F.1.4-4 or by Equation (F.1.4-9).

F.3 Thermal Insulation of Concrete Surface

F.3.1 When the freshly placed concrete encounters a cold wave, concrete surface is prone to cracking due to the larger temperature difference between the interior and exterior and the lower strength of concrete. Therefore, concrete surface insulation is required during the cold wave. During winter season, concrete surface insulation is also required due to the low air temperature, the frequent cold waves, the higher temperature inside the concrete mass and the larger temperature difference between interior and exterior of the concrete. The required equivalent surface heat transfer coefficient after application of the insulation protection and the insulation thickness shall be calculated by the following method.

F.3.2 Concrete surface insulation during cold wave

a The equivalent surface heat transfer coefficient for concrete dissipating heat to single surface subjected to cold waves may be calculated by Equation (F.3.2-1) to Equation (F.3.2-5):

$$\beta = \frac{\lambda_c}{2u} \sqrt{\frac{\pi}{a_c Q}} \tag{F.3.2-1}$$

$$u = 0.944,9\sqrt{b^2 - 0.236,0} - 0.825,9 \tag{F.3.2-2}$$

$$b = \frac{\rho_1 E(\tau_m)\alpha A}{(1-\mu)(\sigma_a - \sigma_0)} \tag{F.3.2-3}$$

$$\rho_1 = \frac{0.830 + 0.051\tau_m}{1 + 0.051\tau_m} e^{-0.095(P-1)^{0.60}} \tag{F.3.2-4}$$

$$\tau_m = \tau_1 + \Delta + \frac{1}{2}P \tag{F.3.2-5}$$

$$P = Q + \Delta$$

$$\Delta = 0.4gQ$$

$$g = \frac{2}{\pi} \tan^{-1}\left(\frac{1}{1 + \frac{1}{u'}}\right)$$

$$u' = \frac{\lambda_c}{2\beta_{eq}} \sqrt{\frac{\pi}{Qa_c}}$$

where a_c —thermal diffusivity of concrete, m²/d;
   λ_c —thermal conductivity of concrete, kJ/(m·h·℃);

ρ_1 —coefficient in consideration of creep effect;

$E(\tau_m)$ —average Young's modulus of concrete during period of cold wave, MPa;

α —coefficient of thermal expansion of concrete;

A —amplitude of air temperature drop, ℃;

μ —Poisson's ratio of concrete;

σ_a —allowable tensile stress of concrete, MPa;

σ_0 —initial stress caused by other factors, MPa;

τ_m —mean age of concrete during cold wave, d;

τ_1 —age of concrete at the beginning of the cold wave, d;

Q —duration of the cold wave, d;

β_{eq} —equivalent surface heat transfer coefficient of concrete surface, kJ/(m² · h · ℃).

The calculation is started from an assumed value of β_{eq}. The value of β can be computed. Then substitute this value into the equations and re-compute again. Repeat this process to obtain the required value of β. Based on such a value, the required insulation thickness will be determined according to performance of the insulation material. For the locations such as edges and corners, etc., where heat can flow concurrently in two or more directions, thickness of the insulation layer shall be increased accordingly.

b Thickness of insulation material may be calculated by Equation (F.3.2-6):

$$h = k_1 k_2 \lambda_s \left(\frac{1}{\beta} - \frac{1}{\beta_0} \right) \quad \text{(F.3.2-6)}$$

Where h —thickness of insulation material, m;

λ_s —thermal conductivity of the insulation material, see Table F.3.2-1, kJ/(m · h · ℃);

β_0 —surface heat transfer coefficient of concrete surface without insulation protection, kJ/(m² · h · ℃);

k_1 —wind speed correction factor, see Table F.3.2-2;

k_2 —moisture correction factor, 3-5 for wet material and 1 for dry material.

Table F.3.2-1 Thermal conductivity of various insulation materials, λ

Name of Material	λ [kJ(m·h·℃)]	Name of Material	λ [kJ(m·h·℃)]
Foam	0.125,6	Expanded perlite	0.167,5
Glass wool blanket	0.167,4	Bitumen	0.938
Wood board	0.837	Dry cotton wool	0.154,9
Saw dusts	0.628	Bituminous felt	0.167
Straw or plait mat	0.502	Dry sand	1.172
Boiler slag	1.674	Wet sand	4.06
Cane fiber board	0.167	Mineral wool	0.209
Asbestos felt	0.419	Hemp felt	0.188
Foamed concrete	0.377	Common paper board	0.628

Table F.3.2-2 Wind speed correction coefficient

Wind permeability of insulation layer		Wind speed up to 4 m/s	Wind speed over 4 m/s
Wind-permeable insulation layer (straw mat & sawdust, etc.)	Without insulation layer	2.6	3.0
	With external wind-proof insulation layer	1.6	1.9
	With internal windproof insulation layer	2.0	2.3
	With both external and internal windproof insulation layer	1.3	1.5
Wind proof insulation layer		1.3	1.5

F.3.3 Concrete surface insulation during winter seasons

Equation (F.3.2-1) may still be used to calculate the equivalent surface heat transfer coefficient of concrete surface during winter season for a unidirectional heat transfer surface, but the value b shall be calculated by Equation (F.3.3-1) as follows:

$$b = \frac{r\rho_2 E(\tau_m)\alpha A}{(1-\mu)(\sigma_a - \sigma_0)} \quad (\text{F.3.3-1})$$

$$\rho_2 = \frac{0.830 + 0.051\tau_m}{1.00 + 0.051\tau_m} e^{-0.104 p^{0.35}} \quad (\text{F.3.3-2})$$

where r —restraint coefficient, related to length of concrete block, referencing Figure F.3.3.

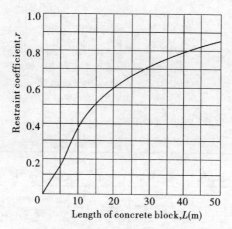

Figure F.3.3 Restraint coefficient, r

F.4 Thermal Stresses

F.4.1 Thermal stresses of base concrete blocks

Calculation of thermal stresses in base concrete is mainly aimed at verification of the horizontal stresses in the central part of the concrete block and the shear stresses along the base plane. The thermal stresses may be calculated by finite element method. Input the calculation results of the above-mentioned temperature field, the boundary conditions and corresponding data according to user instruction of computer program, calculation may be carried out by computer. The thermal stresses may also be calculated by the influence line method, i.e. supposing that the thermal stress of the base concrete block is a single field linear elastic stress problem, the thermal stress (σ_1) induced by the temperature difference between the placing temperature and the final stable temperature, and the thermal stress (σ_2) caused by the drop of temperature induced by hydration heat may be separately calculated, and then they will be superimposed. The stress induced by autogenous volume change may also be taken into account if after verification.

$$\sigma = \sigma_1 + \sigma_2 \qquad (\text{F.4.1-1})$$

a Difference between the placing temperature and the final stable temperature is a uniform temperature field, the stress of which may be calculated by the restraint coefficient method:

$$\sigma_1 = K_p \frac{RE_c \alpha}{1-\mu}(T_p - T_f) \qquad (\text{F.4.1-2})$$

where K_p —stress relaxation factor induced by creep, take 0.5 if no test data is available;

R —degree of restraint of foundation to the concrete block. When Young's modulus of concrete, E_c is as much as that of the foundation rock, E_R, the value of R may be taken from Table F.4.1-1. When the Young's modulus of concrete, E_c is different from that of the foundation rock, E_R, the value of R at the dam-foundation interface may be taken from Table F.4.1-2, and the value of R above the base plane may be reduced proportionally;

E_c—Young's modulus of concrete, MPa;

μ —Poisson's ratio of concrete;

α —coefficient of thermal expansion of concrete;

T_p—placement temperature of concrete, ℃;

T_f—final stable temperature of dam mass, ℃.

Table F.4.1-1 Degree of restraint of foundation

$\frac{y}{L}$	0	0.1	0.2	0.3	0.4	0.5
R	0.61	0.44	0.27	0.16	0.10	0

Notes: 1. y is height of computational point above the contact between dam concrete and foundation, m.
2. L is dimension of the longer side of concrete block, m.

Table F.4.1-2 Degree of restraint of foundation at dam-foundation interface

$\frac{E_R}{E}$	0	0.5	1.0	1.5	2.0	3.0	4.0
R	1.0	0.72	0.61	0.51	0.44	0.36	0.32

b Stress caused by temperature drop σ_2 of the hydration heat may be calculated by the influence line method, for which the envelope of the maximum temperature rise of hydration heat in each lift of the base concrete block may be taken as the computational temperature difference:

$$\sigma_2 = \frac{K_p E_c \alpha}{1-\mu}\left[T(y) - \frac{1}{L}\Sigma A_y(\zeta)T(\zeta)\Delta y\right] \quad (\text{F.4.1-3})$$

where E_c —Young's modulus of concrete, MPa;

$T(y)$ —temperature at computational point y, ℃;

$A_y(\zeta)$ —the normal stress influence coefficient at the computational point y on the influence line, which is obtained by applying a pair of unit load $P=1$ at the point $y=\zeta$, may be obtained from Figure F.4.1-1 and Figure F.4.1-2;

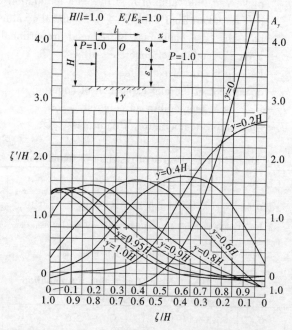

Figure F.4.1-1 Temperature stress influence line of concrete block, $E_c = E_R$

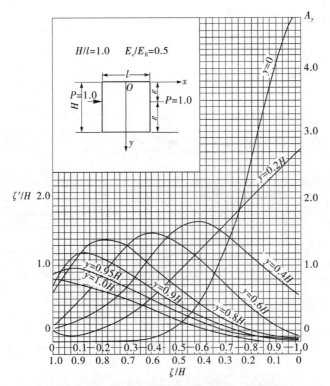

Figure F.4.1-2 Thermal stress influence line of concrete block, $E_c = E_R/2$

$T(\zeta)$ —temperature at the point $y = \zeta$, ℃;

Δy —increment of coordinate y, m;

l —dimension of the longer side of concrete block, m.

F.4.2 Thermal stress at concrete surface

Stress distribution from surface towards the interior in a horizontal section or a vertical section of the concrete block may be calculated either by the finite element method or by the influence line method, based on distribution of the temperature difference of the surface temperature field in various time steps. The equation of surface stress calculation using the influence line method is the same as Equation (F.4.1-3), in which the influence lines are given Figure F.4.2-1 and Figure F.4.2-2. Distribution of the temperature field may be obtained by the finite difference method.

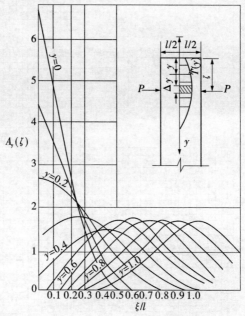

Figure F.4.2-1 Stress influence line of the rectangular section

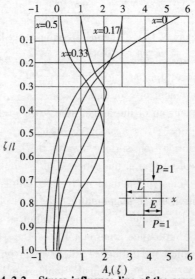

Figure F.4.2-2 Stress influence line of the square section

F. 4. 3 Thermal stress control

a Control of horizontal tensile stress and principal tensile stress for dam body

$$\sigma \leqslant \frac{\varepsilon_p E_c}{K_f} \quad (\text{F.4.3-1})$$

where σ —sum of thermal stresses caused by various temperature differences, MPa;

ε_p —strain capacity of concrete, to be determined by tests for important projects, and $(0.7\text{-}1.0) \times 10^{-4}$ for other projects;

K_f—factor of safety, generally 1.5-2.0, depending on importance of the project and hazard of the cracking.

b Control of vertical tensile stress at horizontal construction joint at the upstream face

$$\sigma \leqslant \frac{R_f C}{K_f} \quad (\text{F.4.3-2})$$

where σ —sum of all vertical tensile stresses at horizontal construction joint near the upstream face, MPa;

R_f—tensile strength of concrete, to be determined by tests for important project;

C —reduction factor of tensile strength at horizontal construction joint, usually 0.6-0.8;

K_f—factor of safety, 1.5-2.0, depending on importance of the project and hazard of the cracking.

Expression of Provisions[1]

The forms shown in Table 1 shall be used to indicate requirements strictly to be followed in order to conform to the document and from which no deviation is permitted.

Table 1　Requirement

Verbal form	Equivalent expressions for use in exceptional cases
Must, shall	Is to Is required to It is required that Has to Only … is permitted It is necessary
Shall not	Is not allowed [permitted] [acceptable] [permissible] Is required to be not Is required that … be not Is not to be

Use "must" for a very strict and mandatory case printed in boldface.
Do not use "may not" instead of "shall not" to express a prohibition.
To express a direct instruction, for example referring to steps to be taken in a test method, use the imperative mood in English.
EXAMPLE　"Switch on the recorder."

The forms shown in Table 2 shall be used to indicate that among several possibilities one is recommended as particularly suitable, without mentioning or excluding others, or that a certain course of action is preferred but not necessarily required, or that (in the negative form) a certain possibility or course of action is deprecated but not prohibited.

---

[1] NOTE: Only singular forms are shown.

Table 2 Recommendation

Verbal form	Equivalent expressions for use in exceptional cases
Should	It is recommended that It is advisable that/to It is suitable that/to It is appropriate that/to It is desirable that/to It is desired that/to It is preferred that/to It is preferable that/to It is favourable that/to It is better that/to
Should not	Corresponding negative forms of expressions above

The forms shown in Table 3 shall be used to indicate a course of action permissible within the limits of the document.

Table 3 Permission

Verbal form	Equivalent expressions for use in exceptional cases
May	Is permitted Is allowed Is permissible
Need not	It is not required that No ··· is required

Do not use "possible" or "impossible" in this context.
Do not use "can" instead of "may" in this context.
NOTE "May" signifies permission expressed by the document, whereas "can" refers to the ability of a user of the document or to a possibility open to him/her.

The forms shown in Table 4 shall be used for statements of possibility and capability, whether material, physical or causal.

Table 4 Possibility and capability

Verbal form	Equivalent expressions for use in exceptional cases
Can	Be able to There is a possibility of It is possible to
Cannot	Be unable to There is no possibility of It is not possible to

NOTE   See Notes 1 to Table 3.